DATE DUE

AMERICAN TRYPANOSOMIASIS

World Class Parasites

VOLUME 7

Volumes in the World Class Parasites book series are written for researchers, students and scholars who enjoy reading about excellent research on problems of global significance. Each volume focuses on a parasite, or group of parasites, that has a major impact on human health, or agricultural productivity, and against which we have no satisfactory defense. The volumes are intended to supplement more formal texts that cover taxonomy, life cycles, morphology, vector distribution, symptoms and treatment. They integrate vector, pathogen and host biology and celebrate the diversity of approach that comprises modern parasitological research.

Series Editors
Samuel J. Black, *University of Massachusetts, Amherst, MA, U.S.A.*
J. Richard Seed, *University of North Carolina, Chapel Hill, NC, U.S.A.*

AMERICAN TRYPANOSOMIASIS

edited by

Kevin M. Tyler
*Northwestern University - The Feinberg School of Medicine,
Chicago, USA*

and

Michael A. Miles
*London School of Hygiene and Tropical Medicine
London, UK*

KLUWER ACADEMIC PUBLISHERS
Boston / Dordrecht / London

Distributors for North, Central and South America:
Kluwer Academic Publishers
101 Philip Drive
Assinippi Park
Norwell, Massachusetts 02061 USA
Telephone (781) 871-6600
Fax (781) 681-9045
E-Mail: kluwer@wkap.com

Distributors for all other countries:
Kluwer Academic Publishers Group
Post Office Box 322
3300 AH Dordrecht, THE NETHERLANDS
Telephone 31 786 576 000
Fax 31 786 576 254
E-Mail: services@wkap.nl

 Electronic Services < http://www.wkap.nl>

Library of Congress Cataloging-in-Publication Data

A C.I.P. Catalogue record for this book is available
from the Library of Congress.

American Trypanosomiasis Edited by Kevin M. Tyler and Michael A. Miles
ISBN 1-4020-7323-2

TABLE OF CONTENTS

PREFACE

Nearly a century ago, Carlos Chagas established the flagellated protozoan *Trypanosoma cruzi* as the causative agent of a disease that now bears his name and identified the triatomine bug as its vector. Today transmission of Chagas Disease is in decline, primarily through the efforts of effective public health initiatives in the Southern Cone of South America. Nevertheless, in some areas transmission is still widespread. Moreover, because of the chronic, life-long nature of the disease, there is a legacy of over 16 million people who still carry the disease. This legacy is not only a human tragedy, but continues to pose a threat to blood transfusions throughout the Americas. It is clear that if transmission is to continue to dwindle, then the available arsenal must be expanded by improvements in therapeutic, diagnostic, vector control and public health strategies which can only be driven from continued scientific research. In this volume we aim to detail the status quo of research in these areas.

Trypanosomes are an evolutionarily ancient organisms which are fascinating to study because they use mechanisms, structures and sometimes whole organelles which are often completely novel in biology to survive. In each such instance lies one or more targets for rational drug design. In mammals, *T. cruzi* has adopted an intracellular strategy to evade the immune system. While not the only pathogen to do so, the mechanisms by which *T. cruzi* enters cells and replicates within them are novel and of intense interest. It may be that by selectively blocking invasion of new cells one can arrest disease pathogenesis and protect against infection. Modern molecular techniques wielded in the context of the recent elucidation of host and parasite genomes mean that research on the parasite itself and on host parasite interactions has made rapid progress in the last few years.

The complex pathologies of Chagas disease vary both with the infecting strain and with the genetic makeup of the person that contracts it. Moreover, pathology can be further complicated by environmental factors such as transient or prolonged immunosuppression as in the case of AIDS or anti-cancer therapy. *T. cruzi* has diverse genetics, encompassed by two primary lineages, but it is not yet clear how the parasites genetic composition relates to observed pathology. Even the pathogenesis of Chagas disease itself is controversial, since it is not known whether autoimmunity, which is detected in patients, is the cause of significant pathology and hence warrants specific treatment.

Available anti-parasite treatment does appear to impact positively on the disease outcomes. Current drugs are however, poorly tolerated by some patients and is not clear how patients is affected by such treatment. The development of new drugs remain a priority for this disease. The advances in therapeutics and vector control reported in this volume are likely to augment control programs in the near future. Taken together then, this volume provides considerable reason for optimism in the future prosecution of Chagas Disease.

Kevin Tyler and Michael Miles

ACKNOWLEDGEMENTS

On behalf of ourselves and the contributing authors, we would like to thank the philanthropic and governmental organizations which continue to support research and control programs towards the eradication of Chagas disease. We gratefully acknowledge the editorial assistance of Manu Davies and Kathryn Buchanan in the preparation of this volume.

THE LIFE CYCLE OF *TRYPANOSOMA CRUZI*

K. M. Tyler, C. L. Olson and D. M. Engman
Departments of Microbiology-Immunology and Pathology
Feinberg Medical School of Northwestern University, Chicago, IL 60611

ABSTRACT

Since the discovery of *Trypanosoma cruzi* as the parasite that causes Chagas disease, nearly a century ago, the details of the organism's life cycle have fascinated scientists. *T. cruzi* is a single-celled eukaryote with a complex life cycle alternating between reduviid bug vectors and vertebrate hosts. It is able to adapt via the process of cellular differentiation to replicate within the diverse environments represented of the insect's gut and host cell cytoplasm. These adaptive transformations take the form of coordinated changes in morphology, metabolism and cell cycle regulation. Different life cycle stages of *T. cruzi* show dramatically different protein and RNA profiles, which are the end result of unusual mechanisms for regulating gene expression. In recent years, new molecular techniques have been brought to bear on the life cycle dramatically increasing our knowledge of the strategies employed by the parasite to ensure its continued survival.

INTRODUCTION

Chagas disease

The etiologic agent of the chronic and often fatal Chagas disease is the American trypanosome, *Trypanosoma cruzi*, a flagellated protozoan of the order Kinetoplastida. The survival of *T. cruzi* is dependent on the successful transmission between, and the colonization of, two radically different environments: the midgut of the reduviid bug vector and the cytoplasm of the mammalian host cell. As is true of all infections, interruption of the pathogen's life cycle will lead to eradication of the disease. Strategies for interrupting the life cycle include minimizing human contact with the insect vector by improving public housing, reducing or eliminating the vector population, or by manipulation of the vector population to make it refractory to *T. cruzi* infection. These strategies would ideally being employed in tandem, together with treatment of infected individuals using curative chemotherapy. In recent years, such strategies have proven effective in South America, dramatically reducing or eliminating natural infection in most areas.

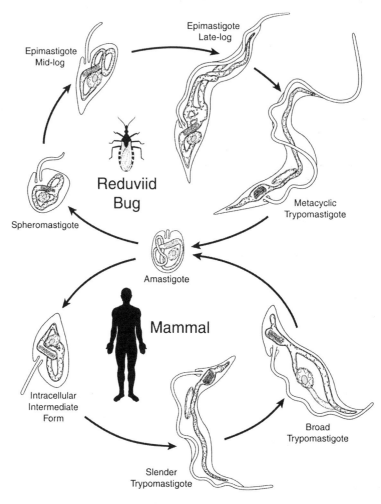

Figure 1. The life cycle of *Trypanosoma cruzi.* See text for details.

Overview of the *T. cruzi* life cycle:

    For the purpose of this discussion we will begin our descriptions of the parasite life cycle with the infection of a mammalian host by metacyclic trypomastigotes present in the excreta of the blood-feeding reduviid bug vector (Figure 1). These are introduced into the host by contamination of the insect bite wound or a variety of mucosal membranes. The non-dividing metacyclic form is able to invade a wide range of phagocytic and non-phagocytic nucleated cells, initially entering a membrane bound (parasitophorous) vacuole. Upon entry, the parasite begins to differentiate to the amastigote form and escapes the vacuole into the cell cytoplasm where the dramatic morphologic transformation, including flagellar involution, is completed. The amastigote re-enters the cell cycle and proliferates until the cell fills with these forms. At this point the amastigotes elongate, reacquiring their long flagella, differentiating to the slender trypomastigote forms via an intracellular epimastigote intermediate. Slender trypomastigotes escaping the cell can invade adjacent cells; alternatively, they can enter the blood and lymph and disseminate, in which case they may begin to differentiate extracellularly. Extracellular differentiation gives rise to the broad trypomastigotes and extracellular amastigotes. A mixture of these three forms may be present in the blood of infected individuals and can be taken up in the blood meal of a reduviid bug. In the bug midgut, remaining trypomastigotes differentiate into amastigotes. As a population, amastigotes first extend their flagella to become spheromastigotes, which then lengthen to become (mid-log) epimastigotes. These epimastigotes continue to elongate as nutrients from the blood meal are exhausted. Finally, after migration to the bug's hindgut, the elongate (late-log) epimastigotes attach to the waxy gut cuticle by their flagella and differentiate into infectious metacyclic trypomastigotes, completing the life cycle.

The trypanosome cell

T. cruzi has the classical features of a eukaryotic cell: membrane bound nucleus, plasma membrane, golgi apparatus and endoplasmic reticulum. However, in common with other members of the Kinetoplastida, *T. cruzi* has several peculiar features, such as a single mitochondrion, the DNA of which lies within a single unit, suborganellar structure - the kinetoplast. The kinetoplast DNA is a linked (catenated) network of hundreds of circular molecules, the minicircles and maxicircles. *T. cruzi* also compartmentalizes glycolysis in membrane bound vesicles called glycosomes, stores minerals in structures known as acidocalcisomes and sequesters membranes in vesicles named reservosomes (see chapter by De Souza, this volume for detailed accounts of the cell biology of this organism).

The cytoskeleton of *T. cruzi* is unusual, in that it is predominantly microtubular with no evidence of microfilament or intermediate filament systems. *T. cruzi* does not possess centrioles. The replicative stages undergo a "closed" mitosis, with a microtubule spindle arising from poorly defined structures in the nuclear membrane. The trypanosome's distinctive morphologies are dictated by a "pellicular" corset of microtubules which closely apposes the plasma membrane.

T. cruzi possesses a single flagellum subtended by a basal body and probasal body which lie within the cell. The basal body is the trypanosome's only defined microtubule organizing center. The flagellum varies in length during the life cycle from over 20 μm to less than 2 μm. The flagellar motor is a ciliary axonemal complex, with the typical 9 + 2 configuration of parallel microtubules. Once the axoneme exits the cell body, it is appended to an unusual semi-crystalline structure called the paraflagellar rod. It is believed that this structure provides support to the flagellar axoneme, increasing its rigidity and playing an essential role in motility. The exterior flagellum is surrounded by a specialized membrane which is rich in sterols and sphingolipids and which contains proteins that do not diffuse into other domains of the surface membrane.

Where the flagellum enters the cell there is a gap in the subpellicular corset, the junction between the pellicular plasma membrane and flagellar membrane at this point takes the form of an invagination known as the flagellar pocket. The majority of vesicular trafficking and nutrient uptake is believed to occur in this area and many receptors localize specifically to this region. A second, smaller invagination proximal to the flagellar pocket, the cytostome, has also been implicated in nutrient uptake.

THE LIFE CYCLE

In the mammalian host

Metacyclic trypomastigotes are able to parasitize a wide range of nucleated mammalian cells. Invasion occurs by one of three distinct mechanisms (Figure 2). The parasite may enter a cell under pressure from its own motility (Figure 2A); this is evidenced by the fact that even lightly fixing cells does not prevent invasion. Nevertheless, this mechanism is thought to be the least important mechanism of invasion. The best-studied entry mechanism is lysosome dependent. In this case, *T. cruzi* organizes the microtubule cytoskeleton of the host cell in order to direct recruitment of lysosomes to the point of parasite attachment. These lysosomes then fuse with the plasma membrane, first forming a junction with the parasite and then creating a vacuolar compartment in which the entering parasite transiently resides (Figure 2B). Finally, invasion may be facilitated by the host actin cytoskeleton. In this case, the parasitophorous vacuole is initially constructed from the plasma membrane of the host cell, which ruffles out along the parasite and encompasses it (Figure 2C). Once within the vacuole by any of these mechanisms, lysosomes continue to traffic to the parasitophorous vacuole. Fusion of lysosomes with the vacuole leads to acidification. This drop in pH serves a dual role in inducing the trypomastigote to differentiate rapidly to an amastigote and also activates a parasite derived porin like molecule - TcTox. This molecule mediates weakening of the membrane of the parasitophorous vacuole and permits escape of the parasite into the cytoplasm.

Flagellar and cell body shortening commences immediately following cell entry; it is a relatively rapid process and precedes rearrangement of the kinetoplast. The first cell division occurs only after a lag period, during this lag period the kinetoplast is rearranged, adopting its replicative morphology (Figure 3). After differentiating to the amastigote form, *T. cruzi* proliferates in the cell cytoplasm. At high density, amastigotes give rise to bloodstream trypomastigotes via a range of intermediate morphologies. The triggers for this differentiation are not yet defined, although glucose limitation and contact interaction are obvious candidates. The intermediate forms of this differentiation contain epimastigote forms. It has been suggested that these forms are a *bona fide,* but transient and somewhat cryptic life cycle stage. In differentiating to the trypomastigote form, the kinetoplast morphology is again a late marker of differentiation with some trypanosomes of clearly trypomastigote morphology possessing a replicative kinetoplast structure whilst still in the host cell.

The mechanisms the parasite has adopted for cell entry, which exploit the host cell trafficking machinery, are tremendously intriguing and apparently novel in biology (Yoshida, this volume). The trypanosome exploits host cellular behaviors that are clearly important, but poorly characterized to date. For this reason, such studies are of great general interest to cell biologists and research on *T. cruzi* invasion has led to great insights into cellular phenomena, such as the role of lysosomal exocytosis in cellular repair. In the past few years, new technologies have greatly facilitated the study of cellular invasion pathways and of trypanosome survival within living cells.

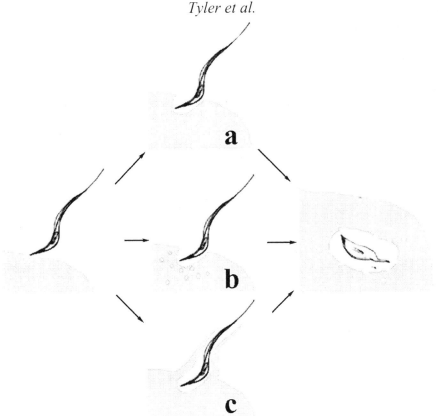

Figure 2. Mechanisms by which *T. cruzi* trypomastigotes can invade nucleated cells. a) **Host cell independent**. The trypanosome attaches to the host cell and, under pressure from its own motility, effects entry into the host cell. b) **Lysosome mediated**. The trypanosome utilizes the host microtubule cytoskeleton to traffic lysosomes to the attachment site, where the extra membrane facilitates entry into a parasitophorous vacuole. c) **Membrane ruffling**. An actin based mechanism akin to phagocytosis in which the cell is stimulated to extend processes along the parasite, facilitating entry. Regardless of mechanism or mechanisms (since they are not mutually exclusive) the endpoint is the parasite resident in a parasitophorous vacuole to which lysosomes traffic, and initiate differentiation and escape into the cytoplasm.

From the mammalian bloodstream to the reduviid digestive tract

Slender trypomastigotes are readily seen escaping from packed pseudocysts (infected cells). It is believed that these slender forms are committed to a program of differentiation to the amastigote form, and that this program will take place whether the parasite is present in the peripheral blood, cytoplasm or reduviid bug gut. The rate of this differentiation is acutely sensitive to pH. In the acid environment of the parasitophorous vacuole, differentiation is rapid, the cell body apparently shortening more rapidly than the flagellum is able, leading to club-shaped intermediate morphologies. At more neutral pH, in the blood, slower kinetics of differentiation are observed and broad (stout or stumpy) forms appear to be the intermediate morphologies *en route* to the amastigote form.

The trypomastigotes in the peripheral blood are pleomorphic, consisting of both slender and broad forms. The ratio of these forms varies from strain to strain, perhaps reflecting differences in the rate of

6 *Tyler et al.*

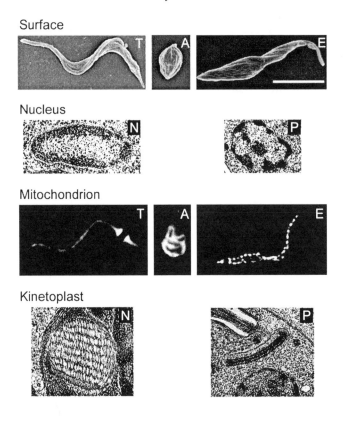

Figure 3. Coordinate changes in morphology and organellar structure during the life cycle. **Surface.** Scanning electron micrographs of the trypomastigote (T) amastigote (A) epimastigote (E). The left-handed helical twist of all forms is dictated by the trypanosome cytoskeleton. SEM images are taken at the same magnification. Scale bar provided is 5 μm. **Nucleus**. Nuclear structure is dramatically different between the non-proliferative (N) trypomastigote and the proliferative (P) amastigote and epimastigote stages. In contrast to the round nuclei of the proliferative stages, no nucleoli are visible in elongate trypomastigote nuclei and heterochromatin is more abundant. **Mitochondrion**. The morphology of the mitochondrial tubule and its complexity change during the life cycle. The linear tubule of the trypomastigote is transformed via the simple geometries of the amastigote and log phase epimastigote chondriome to a complex labyrinth in the stationary phase epimastigote form. **Kinetoplast.** Accompanying changes in the chondriome morphology, the kinetoplast morphology has a dramatically different structure in the trypomastigote forms (N) where it has an open "basket" like appearance compared with the proliferative stages (P) where the kinetoplast has a simple bilamellar appearance in cross-section. Although clearly linked, at a molecular level, the links between the changes in morphology, nuclear, chondriome and kinetoplast structures during the life cycle remain poorly characterized. TEMs kindly provided by Dr. C. Sterling, University of Arizona.

differentiation or in the invasiveness of the trypomastigotes in the different strains studied. When a pleomorphic population of bloodstream trypomastigotes (and up to 10% amastigotes) is ingested by a reduviid during a blood meal, parasites first pass into the bug midgut. Here, the trypomastigotes undergo differentiation to amastigote forms. The amastigote forms (which are generally 3-5 μm in diameter) replicate and differentiate into epimastigotes, which are also able to replicate. Initially, the amastigote-like forms swell, roughly doubling in diameter, and elongate their flagella, which in turn begin to beat visibly. At this stage the forms are sometimes referred to as spheromastigotes. The cell body and flagellum of the spheromastigote elongate, giving rise to the classical epimastigote form which has a varied morphology and which can become quite long (in excess of 30 μm). The transformation from amastigote to elongate epimastigote appears to be reversible and dependent upon the concentration of free monosaccharides in the environment. Since amastigotes, spheromastigotes and epimastigotes are all proliferative forms, the transitions from one of these forms to another appear to lie in a continuum rather than being discrete steps.

The progression from slender trypomastigote to elongate epimastigote represents a progressive change from an environment rich in glucose (host bloodstream) to an environment which is extremely poor in monosaccharides (bug hindgut). In an apparent adaptation to this depletion of simple saccharides, an incremental biogenesis of the mitochondrion is readily observed (Figure 3). The single tubule of the slender trypomastigote mitochondrion splits into two tubules linked at either end in the broad form, which twist together in the amastigote mitochondrion often showing a figure 8-like appearance. As the cell body elongates into the epimastigote form, additional processes arise, but a simple geometry is initially maintained in shorter epimastigotes. As nutrients dwindle, mitochondrial complexity increases until, in the elongate epimastigote, multiple intertwined processes occupy much of the volume of the cell. Accompanying these changes is an increase in the expression of both key electron transport chain components and mitochondrial chaperones. Importantly, cytochrome reductase (complex III) is activated during the trypomastigote to amastigote transition, while dependence on cytochrome oxidase (complex IV), rather than the trypanosome's plant-like alternative oxidase, is developed only during the transition from short (mid-log) to elongate (late-log) epimastigote.

This progressive increase in mitochondrial complexity and activity parallels changes seen in *Trypanosoma brucei* during transition from mammalian host to its tsetse fly, insect vector. As with *T. brucei*, key enzymes associated with amino acid metabolism, the primary source of nutrients in the insect gut, are upregulated in the vector stages. Unlike *T. brucei*, Kreb's cycle components such as dihydrolipoamide dehydrogenase, appear to be constitutively active in all life cycle forms. Moreover, early spectophotometric studies showed that at least some cytochromes are present throughout the life cycle of *T. cruzi*. So it may be that some parts of the electron transport chain are constitutively active through the life cycle, working in concert with the trypanosome alternative oxidase.

The mitochondrial genome of *T. cruzi* encodes subunits of cytochrome reductase and cytochrome oxidase in its maxicircles. To make these subunits functional, the RNA must be transcribed from the maxicircle and amended post-transcriptionally by the addition and occasional deletion of uridine residues. This process, known as RNA editing, is mediated by machinery that includes the guide RNAs encoded by the kinetoplast minicircles. It would be surprising if the kinetoplast DNA (kDNA) encoded subunits were not coordinately regulated with their nuclear encoded counterparts as they are in *T. brucei*. In *T. brucei,* one level of control for mitochondrial encoded gene expression is stage specific editing of constitutively produced transcripts. To date there is no evidence of stage specific RNA editing in *T. cruzi*. The observation of a massive and unexplained structural change in the kDNA of the trypomastigote may, however, indicate an entirely different mechanism for regulation of *T. cruzi* kinetoplast encoded proteins. It may be that transcription of the trypomastigote kinetoplast is essentially shut down when the kDNA adopts the basket like configuration (Figure 3). This would parallel gross morphological changes in the nucleus which have been correlated with pan-regulation of transcription between trypomastigote and replicative forms.

Metacyclogenesis

In the midgut of its triatomine bug vector, *T. cruzi* epimastigotes proliferate in the nutrient rich environment of a recent blood meal. As the meal is digested and the parasite density increases, the environment becomes nutrient poor and epimastigotes become more elongate. Eventually, epimastigotes reaching the insect rectum attach by their flagella and undergo metacyclogenesis to human infective trypomastigote forms. Metacyclogenesis occurs when epimastigotes from the nutrient poor hindgut adhere to the waxy cuticle of the reduviid bug rectum, initiating a dramatic morphological change. Once formed, metacyclics detach from the waxy cuticle and are excreted. Contamination of the reduviid bite wound of the mammalian host with these excreta leads to infection, completing the life cycle.

Metacyclogenesis can be described in two parts, the first enabling the second. First, the trypanosome senses loss of sugars from its environment and responds by activating its mitochondrion and by elongating its cell body and flagellum. The trypanosome flagellar membrane, which is sterol rich and more hydrophobic that the somatic membrane, is thus lengthened. Second, this flagellar lengthening permits the trypanosomes to adhere to a hydrophobic surface and it is this interaction which triggers metacyclogenesis. This trigger for metacyclogenesis is known to be cAMP mediated. Although the machinery involved in transducing the attachment signal has not been demonstrated it is presumably localized, at least in part, in the flagellar membrane. One family of adenylate cyclases has been discovered in trypanosomes and at least some members of this family are known to be resident in the flagellar membrane. Moreover, cAMP-regulated protein kinase (PKA) homolog is also believed to be resident in the flagellum - clearly these are candidates for metacyclogenesis control molecules.

As long as there are sufficient nutrients - particularly the exogenous sugars which must not dip below a critical level necessary for powering the differentiation - the hydrophobic interaction between the flagellum and the substrate to which it attaches is sufficient to trigger the differentiation process. Saccharide limitation and hydrophobic interaction with the flagellum seem necessary and sufficient to drive metacyclogenesis. *In vivo*, however, there is experimental support for the ability of a triatomine factor in haemolymph to induce metacyclogenesis and a role for accumulated parasite factors or excretory products in metacyclogenesis has not been ruled out. Metacyclogenesis involves coordination of an extreme morphological event with arrest in the cell cycle, changes in antigenicity, reduced mitochondrial activity and acquisition of infectivity. The manner in which these events are coordinated remains almost completely unstudied.

TRYPANOSOME DIFFERENTIATION

T. cruzi employs cellular differentiation as a strategy for adapting to the diverse environments represented by its host and vector. Consequently, the study of cellular differentiation in this organism is synonymous with the study of its life cycle. Differentiation is highly controlled, affecting many fundamental processes within the cell. It is characterized by profound changes in cellular morphology, motility and metabolism, typically in response to external stimuli. Differentiation normally involves a change in the pattern of gene expression, such that the gene products impact in a coordinated fashion on the cellular processes involved. One established approach to understanding such a complex phenomenon, in any given cell type, involves first identifying what is changing during the course of differentiation and then determining the mechanisms by which such change is brought about.

During life cycle transitions *T. cruzi* is known to regulate several key areas of its cell biology:

1) Cell surface - allowing the parasite to interact productively with successive environments
2) Cytoskeleton - both pellicular and flagellar, affecting morphology and motility
3) Nutrient uptake and metabolism - notably mitochondrial structure and activity and the presence or absence of a cytostome, but also regulation of some glycosomal and cytosolic enzymes
4) Cell cycle - with the invasive forms being non-proliferative and reentering the cell cycle upon reaching a stable environment
5) Defense - since the immune responses of the human host and bug gut are immensely different, different molecules are required by the trypanosome to survive them.

The signal to differentiate

Cellular differentiation is a cascade linking the impetus to differentiate with the multiple effects of differentiation. Signaling cascades are the focus of intense research in *T. cruzi* research, in both the parasite (DoCampo, this volume) and the parasitized host cell (Yoshida, this volume).

The functions of cAMP and inositol metabolites are often antagonistic and it is clear that the regulation of cAMP levels and inositide metabolism is critical to the control of *T. cruzi* differentiation. A role for cAMP in differentiation from dividing epimastigotes to non-dividing metacyclic forms has been directly demonstrated using lipophilic analogs of cAMP which induce metacyclogenesis directly. In contrast, inositide metabolism has been shown not to affect metacyclogenesis, but rather to promote trypanosome proliferation and to be critically involved (through phospholipase C) in differentiation from the trypomastigote to amastigote form. This differentiation involves re-entry to the proliferative cell cycle and has also been shown to be accelerated by inhibition of type I protein phosphatases. Interestingly, signaling proteins are increasingly being localized to the sterol rich environment of the flagellar membrane and to the flagellar cytoskeleton. Notably, these proteins include calcium binding proteins and adenylate cyclases, which has led to the idea of the flagellum being regarded as an organelle not just for motility and adhesion but also as a sensory center. It is likely that components of the signaling relays regulating flagellar beat and differentiation are both residents of lipid rafts in the trypanosome flagellum.

Modulation of gene expression in *T. cruzi*

Once the stimulus to differentiate has been transduced into the cell, the machinery responding to the call for differentiation falls readily into two parts: the manufacture of new biological molecules and assemblies, and the breakdown of old ones.

Stage specific protein turnover

Proteins targeted for destruction can be tagged (by ubiquitin ligases) with ubiquitin. Once tagged, these molecules are selectively degraded by large assemblies know as proteosomes. Recent evidence using chemical inhibitors of proteosomes suggest that life cycle differentiations in *T. cruzi* are dependent on proteosomal degradation. Support for this view comes from *T. brucei,* where two molecules with ubiquitin ligase homology have been shown to participate directly in control of morphological change during differentiation. It is also important to note that cell cycle control is modulated by proteosomal degradation of ubiquitinated histones, emphasizing a link between cell cycle and morphology in trypanosomes. Although the proteosome is undoubtedly a key player in differentiation, other proteases such as the major cysteine protease of *T. cruzi* (cruzipain/cruzain) have also been suggested to play an enabling role in *T. cruzi* differentiation, although its precise role and targets have not yet been discovered.

Control of new protein synthesis

The polymerase II of *T. cruzi* transcribes coding sequences as large polycistronic units and, to date, promoters or transcriptional start sites have proven difficult to detect. This is in contrast to most metazoa, in which transcription initiation is a major point of regulation. Consequently, it appears that *T. cruzi* modulates protein synthesis during its life cycle by controlling RNA maturation and by employing a range of post-transcriptional mechanisms.

In the monocistronic transcription of metazoan protein encoding genes, the pre-mRNA is processed by addition of a poly (A) tail to the 3' end and by co-transcriptional capping of the 5' end. These modifications confer stability to the mature mRNA and allow its recognition by the ribosomes for translation. Polycistronic transcripts from trypanosomes cannot be capped co-transcriptionally. Instead, a stabilizing 5' terminus (spliced leader sequence) with a cap is added to all protein-coding mRNAs by the process of trans-splicing. Generally, trans-splicing of the 5' end of one trypansome mRNA is linked directly to the polyadenylation of the 3' end of the upstream mRNA.

The mRNAs of *T. cruzi* stage regulated proteins may show entirely different patterns of mRNA abundance than mRNAs encoded by adjacent genes, even though both genes are constitutively transcribed as the same polycistronic transcript. One such example is the amastigote specific amastin protein that is encoded by genes in alternating tandem array with the constitutively expressed tuzin protein. The profile of mRNA abundance for these two genes is dramatically different through the life cycle. In this case and in most others that have been investigated, *T. cruzi* mRNA abundance is dictated by the longevity of the mRNA and this, in turn, is dictated by sequences within the 3' untranslated region (UTR). These UTR sequences are thought to specifically bind regulatory proteins that can either stabilize or destabilize them. The first such UTR binding proteins have now been discovered and putatively linked to mRNA stability. Other co-transcriptional and post-transcriptional mechanisms may also play a role in determining mRNA abundance, but are less well worked out. Sometimes protein expression does not correlate with RNA levels in trypanosomes; in such cases control is thought to lie at the level of translation. Finally, it has been observed that both the nucleus (chromatin) and kinetoplast (kDNA) structures vary dramatically during the life cycle, between replicative and non replicative forms, affecting not only their overall morphology, but also accessibility of their DNAs to digestive (and other) enzymes (Figure 3). It has been suggested that the structural changes observed correlate directly with a net down regulation of transcription in the non-proliferative forms. To date, however, it is not clear whether direct chemical modification of the DNA has a role to play as it does in transcriptional silencing of metazoan cells.

Further reading
Brener Z. 1973. Biology of *Trypanosoma cruzi*. Annu Rev Microbiol 27:347-82.
Burleigh BA, Andrews NW. 1998. Signaling and host cell invasion by *Trypanosoma cruzi*. Curr Opin Microbiol 1:461-5.
De Souza W. 1984. Cell biology of *Trypanosoma cruzi*. Int Rev Cytol 86:197-283.
Gull K. 1999. The cytoskeleton of trypanosomatid parasites. Annu Rev Microbiol 53:629-55.
Hecker H, Betschart B, Bender K, Burri M, Schlimme W. 1994. The chromatin of trypanosomes. Int J Parasitol 24:809-19.
Matthews KR. 1999. Developments in the differentiation of *Trypanosoma brucei*. Parasitol Today 15:76-80.
Naula C, Seebeck T. 2000. Cyclic AMP signaling in trypanosomatids. Parasitol Today 16:35-8.
Parsons M, Ruben L. 2000. Pathways involved in environmental sensing in trypanosomatids. Parasitol Today 16:56-62.
Priest JW, Hajduk SL. 1994. Developmental regulation of mitochondrial biogenesis in *Trypanosoma brucei*. J Bioenerg Biomembr 26:179-91.
Kollien AH, Schaub GA. 2000. The development of *Trypanosoma cruzi* in Triatominae. Parasitol Today 16:381-7.
Elias MC, Marques-Porto R, Freymuller E, Schenkman S. 2001. Transcription rate modulation through the *Trypanosoma cruzi* life cycle occurs in parallel with changes in nuclear organisation. Mol Biochem Parasitol 112:79-90.
Shapiro TA, Englund PT. 1995. The structure and replication of kinetoplast DNA. Annu Rev Microbiol 49:117-43.
Aikawa M, Sterling CR. 1974. Intacellular parasitic protozoa. Academic Press.
Tyler KM, Engman DM. 2001. The life cycle of *Trypanosoma cruzi* revisited. Int J Parasitol 31:472-81.
Vickerman K. 1985. Developmental cycles and biology of pathogenic trypanosomes. Br Med Bull 41:105-14.

NOVEL CELL BIOLOGY OF *TRYPANOSOMA CRUZI*

W. de Souza
Laboratório de Ultraestrutura Celular, Instituto de Biofísica Carlos Chagas Filho, Universidade Federal do Rio de Janeiro, 21941-900, CCS, Bloco G, Ilha do Fundão, Rio de Janeiro, RJ, Brazil

ABSTRACT

This paper reviews basic aspects of the cell biology of trypanosomatids with a special emphasis on results obtained with *Trypanosoma cruzi*. The following structures are discussed: (a) the various domains of the cell surface that can be identified using freeze-fracture and cytochemistry, and the association of the plasma membrane with microtubules; (b) the organization of the flagellum, especially the paraflagellar rod; (c) the structure of the nucleus and its behaviour during cell division; (d) the kinetoplast and the replication of the kinetoplast DNA; (e) the organization of the glycosome and the import of its proteins; (f) the endocytic pathway, emphasizing the role played by the cytostome and the reservosomes; (g) the organization, distribution and function of the acidocalcisomes.

INTRODUCTION

In addition to being responsible for diseases of widespread interest, protozoa of the Family Trypanosomatidae are also of biological interest since they possess specialized organelles and structures. We will briefly review here some of the most important features of these protozoa. In general, what will be described here is valid for all trypanosomatids. We will, however, emphasize those results obtained using *Trypanosoma cruzi*.

THE CELL SURFACE AND SURFACE DOMAINS

The cell body of all evolutive forms in the *T. cruzi* life cycle is enveloped by a typical plasma membrane. Although the appearance of this membrane is the same for all regions of the protozoan, as seen by transmission electron microscopy of thin sections, other techniques have revealed the existence of marked differences in the membrane organization, which varies according to the region. Indeed, we can state that there are at least four macrodomains of the surface of *T. cruzi*, which can be easily recognized based on morphology, on organization of intramembranous particles seen in freeze-fracture replicas and on association with sub-pellicular microtubules (Reviews in de Souza, 1989; 1995). The membrane lining the flagellum represents one domain. A second domain corresponds to the membrane lining the flagellar pocket region. The third domain corresponds to the membrane lining most of the cell body. The fourth domain is observed in the membrane lining the cytostome found in amastigote and epimastigote forms. In addition to the macrodomains, special microdomains have been identified: The region of attachment of the flagellum to the cell body, the base of the flagellum, etc. These domains need to be better characterized biochemically, since some markers such as Ca^{2+}-binding proteins and a glucose transporter are known to be concentrated in the flagellar membrane (Review in Landfear and Ignatuschenko, 2001).

14   *De Souza*

The observation of the actual surface of *T. cruzi* using the fracture-flip technique also revealed differences in the exposition of membrane-associated macromolecules according to the protozoan region and the developmental stage. The surface of epimastigotes is relatively smooth, except at the region of the cytostome (Figure 1). In contrast, the surface of trypomastigote forms is rugous, with many particles protruding toward the outer face (Figure 2). These protrusions may correspond to macromolecules exposed on the parasite surface and associated with the plasma membrane via a GPI anchor (Review in de Souza, 1995).

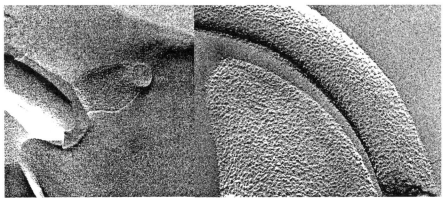

Figures 1 and 2. General views of the actual surface of epimastigote and trypomastigote forms of T. cruzi, respectively. The surface of trypomastigotes is much more rugous. Rugosity of the region of the cytostome is also evident (asterisk) (After Pimenta et al., 1989)

THE CYTOSKELETON

One of the characteristic features of the Trypanosomatidae is the presence of a layer of microtubules localized below the plasma membrane and designated as subpellicular microtubules. It has been observed that the microtubules are connected to each other and to the plasma membrane by short filaments, of still unknown nature (Figure 3). This association is probably responsible for the rigidity of the cell and the difficulties found in the disruption of the cell by mechanical means. Transversal sections through different regions of the trypomastigote form of *T. cruzi* show regular spacing of 44 nm between the microtubules (center to center). Occasionally, one or more microtubules may be lacking, especially in the region of attachment of the flagellum to the protozoan body. It was observed that the number of microtubules is related to the diameter of the cell. At the posterior and anterior regions of the protozoan, the number of microtubules is smaller while the largest number is found between the nucleus and the kinetoplast, at the region where the Golgi complex is located. This indicated that the microtubules stop and start at variable positions. It has been shown that new microtubules are inserted between old ones during the cell cycle. Cross-sections of the subpellicular microtubules of *T. cruzi* show that they are hollow structures. Their wall is composed by 13 protofilaments, has a thickness of about 5 nm and the central portion shows a diameter of about 20 nm (Review in Gull, 2001).

Important information has been obtained in recent years on the biochemistry of the sub-pellicular microtubules of trypanosomatids. It has been shown that the microtubules have a defined polarity with plus end oriented towards the posterior region of the cell. Microtubules are essentialy made of tubulins. Alpha and beta tubulins constitute the main subunits forming the microtubule wall. Gamma tubulin, which is a minor component, is involved in microtubule nucleation. Other tubulins such as delta, epsilon and zeta were recently identified in *T. brucei* (Review in Gull, 2001). In addition to the sub-pellicular microtubules, there is a set of four microtubules that originates close to the basal bodies, runs around the flagellar pocket and seems to be involved with the points of attachment of the flagellum. They are more resistant to depolymerization than the sub-pellicular microtubules.

There is little information about filaments (microfilaments and intermediate filaments) in trypanosomatids. Short filaments (6 nm thick) connect the subpellicular microtubules with each other and with the plasma membrane. In some trypanosomatids, a microfibrillar structure has been observed in the region of attachment of the flagellum to the cell body or to the wall of the intestinal tube of the invertebrate host. Microfilaments were never observed in the cytoplasm of *T. cruzi*. However, cytochalasin B, a drug which among other actions interferes with actin microfilaments, induces changes in the morphology of bloodstream trypomastigotes and inhibits movement. The cytochalasin B effect is readily reversed by washing the cells (de Souza, unpublished observations). Using anti-actin antibodies, actin was observed by immunofluorescence microscopy in the cell body and in the flagellum of *T. cruzi*. However, with the resolution provided by the light microscope it was not possible to identify the structures interacting with the antibodies. Actin has been biochemically detected in trypanosomatids but had an unusual DNAse-I binding behavior when compared with actin from higher eukaryotes.

Adjacent to the axoneme of the flagellum of trypanosomatids there is a filamentous, lattice-like structure that has been called the paraflagellar (or paraxial) rod (PFR) (Figure 4). Three different zones of the PFR can be identified: the proximal and distal zones are made of filaments with a mean diameter of 7-10 nm, which intersect at an angle of $100°$. The intermediate zone is made of thinner filaments (about 5 nm) intersecting at an angle of $45°$. The proximal zone is attached to doublets 4 to 7 of the axoneme through special filaments. Biochemical analyses have shown that the PFR is made of two major and several minor proteins. The major proteins are of around 68 and 76 kDa and the analysis of their genes showed extensive similarity between them. Recent studies, using the gene knockout approach, provided clear evidence that the PFR play some role in the protozoan motility. The mutants obtained were either significantly less motile, as observed in *L. mexicana*; or paralyzed, as observed in *T. brucei* (Review in Gull, 2001). It is important to point out, however, that trypanosomatids harboring endosymbionts do not possess a PFR but are highly motile.

A significant portion of the flagellum of *T. cruzi* is attached to the cell body. Observations of thin sections of the attachment zone by electron microscopy show that the junctional complex is formed by a linear series of apposed macular densities, each measuring 25 nm in diameter and formed by an amorphous material spaced at intervals of 90 nm. Freeze-fracture studies indicate that there is a specialization of the flagellar and cell body membrane at this region, showing clusters of intramembranous particles spaced at more or less regular intervals. In *T. cruzi* epimastigotes, a second specialization was found in both faces of the flagellar membrane, which appeared as a linear array of particles, longitudinally oriented in relation to the main axis of the flagellum. The functional role of this structure was not yet determined. Although there are no conclusive studies establishing the nature of the proteins involved in the attachment of the flagellum to the cell body of *T. cruzi*, two groups of proteins deserve some comments. First is a glycoprotein known as gp72. In null mutants of this protein, the flagellum is no longer attached to the cell body (Cooper et al., 1993). Second, several proteins of high molecular weight have been described in trypanosomatids that localized at the attachment region.

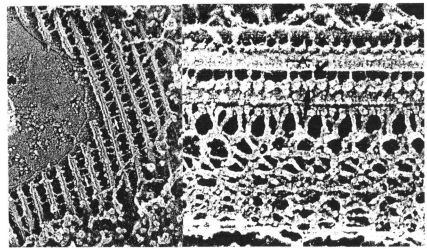

Figures 3 and 4. Deep-etch views of sub-pellicular microtubules and the complex array of filaments which make the paraflagellar rod (After Souto-Padron et al., 1984).

THE NUCLEUS AND PARASITE DIVISION

The nucleus of *T. cruzi* (and other trypanosomatids) is small, measuring about 2.5 μm, with structural organization similar to nuclei of other eukaryotic cells. In *T. cruzi* trypomastigotes, the nucleus is elongated and localized in the central portion of the cell. In spheromastigotes and epimastigotes it has a rounded shape. The nucleus has a typical nuclear membrane complete with pores. In interphasic *T. cruzi* cells, the chromatic material agglomerates into masses at the periphery of the nucleus, below its inner membrane. Occasionally these masses are also found in the more central region. The chromosomes are difficult to distinguish because they do not condense at any stage of the life cycle. However, with the advent of the pulsed field electrophoresis, chromosomes have been identified. The genome of *T. cruzi* consists of 87 Mb of DNA distributed among 30-40 chromosomal bands ranging from 0.45 to 4.0 Mb. See:
http://www.dbbm.fiocruz.br/TcruziDB.

In the center of the nucleus, or situated slightly eccentrically, the nucleolus may be found. During division, changes occur in the organization of the nuclear material. The nucleolus is dispersed during division, reappearing at the final phases of the cell division. In the beginning of the division process, when the basal body is replicating, the first signs of division can be observed in the nucleus. The chromatin material localized below the inner nuclear membrane and the nucleolus disappear. Both are dispersed over the whole nucleus giving it a homogeneous aspect. Immediately after replication of the basal body, when the kinetoplast shows no morphological signs of division, microtubules appear inside the nucleus of trypanosomatids. The nucleus, which has a spherical form, changes into a more oval one with the major axis perpendicular to the direction of the flagellum. During the continuation of the process, the kinetoplast divides; the two newly formed structures move to the sides, and the nucleus becomes progressively elongated.

In all studies made on the nucleus in division, small electron-dense plaques were observed among the intranuclear microtubules (Solari, 1980). In the case of *T. cruzi*, the spindle microtubules were seen in connection with the plaques. A three-dimensional reconstruction of dividing cells showed that the equatorial spindle is formed by about 120 microtubules arranged in two sets of about 60 microtubules running from each pole to the dense plaques and divided into discrete bundles which reach a single plaque. Ten plaques, which were not located in one plane but were distributed within a region which covers about 0.4 µm from both sides of the actual equatorial plane, were identified. The average dimensions of each plaque were: 200 nm length, 120 nm width, and 70 nm of thickness. Each plaque had a symmetrical structure formed by transverse bands. Before nuclear elongation occurs the dense plaques split in two halves and begin to migrate to the polar regions. At this time no microtubules were seen between the two halves of each plaque. All microtubules were localized between the plaques and the poles of the nucleus. In the elongated nucleus, it was possible to see 10 half-plaques on each side of the dividing nucleus. The nature and functional role of the dense plaques are not yet clear. It has been suggested that they could represent specialized parts of noncondensed chromosomes. They could also represent kinetochore-like structures which would play an important role in the process of separation of the nuclear material into the two new cells.

When the nucleus is elongated it divides by constriction in the middle. When the division is completed the chromatin material and the nucleolus reorganize and assume the position seen in interphase cells and the intranuclear microtubules disappear. During the whole process of division the nuclear membrane remains intact, though it does show a more irregular fold. No centrioles have been observed in connection with these microtubules, nor have other structures been found which would suggest participation in their formation. The results obtained until now do not permit a precise statement of the functional role of the intranuclear microtubules in the dividing nucleus. It seems likely, though, that their role is functionally analogous to the role of nuclear microtubules in metazoa, i.e., helping to separate and move the genetic material into the two newly forming nuclei. They may also be engaged in elongating the dividing nucleus. Recent studies correlate the lack of a nucleolus in trypomastigote forms with a decrease in the transcription rates by RNA polymerases I and II when epimastigotes transform into trypomastigotes.

THE KINETOPLAST

T. cruzi, as well as other members of the Trypanosomatidae family, possesses only one mitochondrion which extends throughout the cell body. At a certain portion of the mitochondrion, localized near the basal body, there is a complex array of DNA fibrils in the mitochondrial matrix which forms the structure known as kinetoplast (Reviews in Englund et al., 1996; Morris et al., 2001). This structure is connected to the basal bodies through filamentous structures. The kinetoplast of epimastigote and spheromastigote forms of *T. cruzi* presents a similar morphology. The filamentous material is arranged in a tightly packed row of fibers oriented parallel to the longitudinal axis of the protozoan. The whole structure appears as a slightly concave disk of 1.0 μm in length, and a depth of 0.1 μm. There is a space between the kinetoplast and the inner mitochondrial membrane in which mitochondrial cristae can be seen. At least some fibrils of the kinetoplast DNA (kDNA) make contact with the inner mitochondrial membrane. Such contact is observed mainly in the portion of the mitochondrion that faces the basal body. In epimastigotes and spheromastigotes the kinetoplast may have the appearance of a double layer. In trypomastigotes, however, the kinetoplast presents a spherical shape and kDNA is not so tightly packed.

The kinetoplast represents between 20 and 25% of the total DNA of epimastigotes of *T. cruzi*. Electron microscopy showed that the kinetoplast consists of a network of 20,000 to 30,000 minicircle molecules associated with each other and about 30 maxicircles. Each minicircle has a length of about 0.45 μm, which corresponds to about 1440 base pairs and a molecular weight of 0.94×10^6 Daltons. It has been shown that the diameter of the minicircles from the kDNA isolated from epimastigotes of *T. cruzi* corresponds to the width of the kinetoplast, as seen in thin sections. It was suggested that the minicircles, figure eights, and loops of the long molecules are aligned *in situ* in package fashion to form the kDNA quaternary structure. Biochemical studies indicate the existence of proteins associated with the kinetoplast. The following proteins have been detected in association with the kDNA: type II topoisomerase, DNA polymerase β, minicircle origin-binding protein, kDNA condensing proteins, kinetoplast-associated protein, a nicking enzyme, primase, mitochondrial heat shock proteins, and a set of kinetoplast binding proteins which could bind both mini and maxicircles. Restriction enzyme analysis has shown that the minicircles are heterogeneous in sequence. The maxicircles, with a diameter of about 10 μm, have a size comparable to that found in the mitochondrial DNA of other eukaryotic cells and contains a DNA sequence that encodes some of the proteins required for mitochondrial bioenergetic processes. Some of the RNA transcripts of the maxicircles are post-transcriptionally modified in a process known as RNA editing where there is insertion or deletion of uridine residues to form functional open reading frames. Editing specificity is due to guide RNAs encoded by the minicircles.

The division of the kinetoplast is coordinated with the process of cell division. Division in *T. cruzi* begins with the replication of the basal body and flagellum followed by the division of the kinetoplast. In dividing forms the length of the kinetoplast increases until it reaches a certain value when partition occurs in the middle. By using electron microscope autoradiography it has been shown that the incorporation of H^3-thymidine occurred mainly along the periphery of the kinetoplast of dividing culture forms. However, incorporation of material was occasionally found in other regions of the kinetoplast. Studies carried out using fluorescence techniques have shown that (a) kDNA replication takes place nearly concurrent with the nucleus S phase, (b) before replication the minicircles which are covalently closed are released vectorially from the network face near the flagellum, (c) replication initiates in the region localized between the flagellar face of the disk and the mitochondrial membrane and at sites where the universal minicircle sequence binding protein is located, (d) the replicating minicircles then move to two antipodal sites that flank the network. (Review in Morris et al., 2001).

THE GLYCOSOME

Membrane-bound cytoplasmic structures resembling those initially designated as microbodies and late on as peroxisomes in mammalian cells have been described in trypanosomatids since the initial studies on their fine structure. Glycosome diameters are approximately 0.7 μm, they are randomly distributed throughout the cell and appear to have a homogenous and slightly dense matrix. Peroxisomes have been defined as organelles that are bounded by a single membrane and that contain catalase and H_2O_2-producing oxidases. Catalase has been used as an enzyme marker, easily localized using the alkaline diaminobenzidine medium to identify an organelle as peroxisome. The application of this approach in trypanosomatids led to the identification of two groups. One includes digenetic trypanosomatids such as *T. brucei, T. cruzi* and *Leishmania* where no significant enzyme activity could be detected both by enzyme assay and cytochemistry. The other includes the monogenetic trypanosomatids such as *Crithidia, Leptomonas*, etc, where the enzyme is easily detected.

A major contribution to the role played by the peroxisome-like vesicles in trypanosomatids came from the work of Opperdoes and co-workers initially on *T. brucei* and then extended to other members of the family Trypanosomatidae (Reviews in Opperdoes, 1988; Parsons et al., 2001). They found that, using cell fractionation and biochemical studies, the glycolytic enzymes involved in the conversion of glucose to 3-phosphoglycerate are located in the peroxisomes. Based on these results the term glycosome was suggested to designate the microbodies or peroxisomes of trypanosomatids. An interesting observation was that the glycolytic enzymes isolated from the glycosomes present a higher isoelectric point as compared with similar enzymes found in the cytosol of mammalian cells. This explains the intense labeling of glycosomes when trypanosomatids are stained with ethanolic phosphotungstic acid under conditions where basic proteins are localized. In addition to catalase, the peroxisomes of mammalian cells possess more than 50 different enzymes involved in metabolic pathways such as peroxide metabolism, β-oxidation of fatty acids, ether phospholipid synthesis, etc. In addition to the metabolic pathways traditionally associated with the peroxisome, the glycosome (uniquely in trypanosomatids) compartmentalizes other metabolic pathways such as carbon dioxide fixation, purine salvage, and pyrimidine biosynthesis *de novo*.

The glycosome does not possess a genome. Therefore, all proteins found in the glycosome are encoded by nuclear genes, translated on free ribosomes, and post-translationally imported into the organelle. The uptake of proteins into the glycosomes occurs within 5 minutes of protein synthesis. Several types of peroxisomal targeting signals have been identified in mammalian cells (Review in Parsons et al., 2001). One type is a C-terminal Ser-Lys-Leu (SKL), also known as peroxisome targeting sequence type I (PTS1), that can also direct heterologous proteins into the glycosome. Peroxisomes of several species, including the glycosomes of trypanosomatids, are labeled when antibodies recognizing SKL is used. However, there is much evidence that considerably more variation in the signal is tolerated for efficient glycosomal targeting in trypanosomatids than in peroxisomes of mammalian cells. For instance studies that analysed the target of phosphoglycerate kinase showed that a basic amino acid is not required and that the first amino acid of the COOH-terminal peptide can be serine, alanine or cysteine and the third leucine can be replaced by tyrosine or methionine without much effect on glycosomal targeting. It is clear from the results obtained that SKL-like signals are not the only possible targeting signal. Other entry signals are (a) the 39-amino acid extension at the COOH terminus of the glycosomal phosphoglycerate kinase of *Crithidia fasciculata*, (b) a possible N-terminal or internal peptide for the targeting of fructose biphosphate aldolase in *T. brucei*. Flaspohler et al. (1997) used a positive genetic selection procedure to isolate a *Leishmania* mutant with a defect in the import of glycosomal proteins. The mutant, designated as gim1 (glycosome import mutant), mis-localizes to the cytosol a subset of proteins with the SKL-like signal. Phenotypic changes were observed in the mutant, the most interesting being the decrease in the number of lipid inclusions, which may be related to the role played by the glycosomes in the lipid metabolism. The product of the gim1 gene shares substantial amino acid sequence similarity to the peroxin 2 family of proteins of the membrane of peroxisomes and which are required for the biogenesis of the organelle. This observation, in association to that showing similarities in the protein target signals, support the view that peroxisomes and glycosomes have a common evolutionary origin (Review in Parsons et al., 2001).

ENDOCYTIC PATHWAY AND THE RESERVOSOME

All trypanosomatids present a region known as the flagellar pocket, which appears as a depression found in the region of the cell from where the flagellum emerges. It is formed as an invagination of the plasma membrane, which establishes a direct continuity with the membrane of the flagellum. Due to the fact that the membranes lining the cell body and the flagellum establish a physical contact to each other at the point of emergence of the flagellum, the pocket can be considered as a special extracellular compartment in some way isolated from the extracellular medium. Indeed, the use of cytochemical labels has shown that macromolecules and large particles added to the culture medium may be excluded from the pocket. There are several lines of evidence showing that the flagellar pocket is a highly specialized region of the surface of trypanosomatids: (a) it is the only region which does not abutt the layer of sub-pellicular microtubules (b) the membrane lining the pocket differs both from the membrane lining the cell body and the flagellar membrane in terms of distribution of intramembranous particles, and in localization of proteins, including some enzymes; (c) there is much morphological and cytochemical evidence showing that the pocket is the place where intense endocytic and exocytic activity takes place (Reviewed in Landfear and Ignatushenko, 2001).

In the epimastigote and amastigote forms of members of the subgenus *Schizotrypanum*, there is a highly specialized structure known as the cytostome-cytopharynx complex, which is localized close to and in continuity with the flagellar pocket (Porto-Carrero et al., 2000). It appears as a funnel-shaped structure, formed due to a deep invagination of the plasma membrane which may reach the nuclear region (Figure 5). The opening of this complex, which is known as the cytostome, may reach a diameter of 0.3 μm and is significantly smaller in the deeper portion of the cytopharynx. The sub-pellicular microtubules follow the invagination of the plasma membrane. There is a specialized region of the membrane lining the parasite that starts in the opening of the cytostome and projects towards the flagellar pocket region

(Figure 1). Freeze-fracture studies have shown that this area is delimited from the other portions of the plasma membrane by a pallisade-like array of closely associated particles (Review in de Souza, 1995). It has been shown that when epimastigotes are incubated in the presence of gold-labeled macromolecules such as transferrin, LDL, etc., they initially bind to the cytostome region and then are internalized via endocytic vesicles, which are formed at the bottom of the cytopharynx. Following binding to the cytostome macromolecules are rapidly internalized via the cytopharynx and appear in small endocytic vesicles, which bud, primarily from the deeper region of this structure. Subsequently these vesicles fuse to each other to form tubular structures that can be observed in the most central portion of the protozoan.

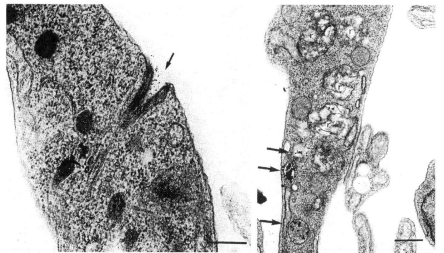

Figures 5 and 6. Binding of gold-labeled transferrin to the initial portion of the cytostome and its subsequent concentration in the reservosomes of epimastigote forms of T. cruzi (After Porto-Carreiro et al., 2001).

Later on the macromolecules are concentrated in structures known as the reservosome (Figure 6). Each epimastigote form presents several reservosomes, mainly localized in the posterior region of the cell, with a mean diameter of 0.7 μm, and surrounded by a unit membrane. The matrix of the reservosome is slightly dense and contains proteins and lipids, which may form some inclusions. The organelle was designated as reservosome based on two criteria. First, because all macromolecules ingested by the parasite through an endocytic process, accumulate in the organelle. Second, because it gradually disappears when epimastigotes are incubated in a poor culture medium, conditions which trigger the process of transformation of non infective epimastigote into infective trypomastigote forms. Most of the endocytic compartments of *T. cruzi* are acidic as indicated by labeling with acridine orange. The determination of the pH of the organelle using the DAMP technique indicated a value of pH 6.0 thus suggesting that the reservosome corresponds to a pre-lysosomal compartment (Soares et al., 1992). No acid phosphatase activity can be systematically detected in the organelle. One characteristic feature of the reservosome in *T. cruzi* is to accumulate a large amount of cruzipain, the major cysteine proteinase found in the cell.

THE ACIDOCALCISOME

Structures designated as electron-dense granules, volutin granules or inclusion vacuoles have been considered for many years as part of the structure of trypanosomatids, but without any special relevance for the cell physiology. The use of X-ray microanalysis, in which it is possible to correlate a cell's ultrastucture with its elemental composition, showed the presence of Ca, Mg, Na, Zn and Fe in the cytoplasmic electron-dense granules of trypanosomatids. A major breakthrough on the study of cytoplasmic vacuoles of trypanosomatids took place in 1994 when Vercesi et al. put together two basic observations: (a) The existence of an intracellular Ca^{2+} pool that was released when the cells were treated with nigericin and the cytoplasm became acidic, and (b) the existence of acidic organelles named as acidosomes (in the slime mold *Dictyostelium discoideum* and in mammals), which possessed both an ATP-driven Ca^{2+}-H^+ antiport and a vacuolar-type H^+-ATPase. In addition, Vercesi et al. noted that Ca^{2+} was released from the intracellular pool not because of acidification of the cytosol by nigericin, but because this drug released the ions from intracellular acidic vacuoles. Based on the fact that the cytoplasmic vacuoles contained a very high Ca^{2+} concentration and a Ca^{2+}-H^+ translocating ATPase activity the organelle, was designated as **acidocalcisome**.

Morphologically, the acidocalcisome appears as a membrane bound structure with an electron-dense content (Figure 7). In routine procedures part of the dense material may be removed, leaving a thin dense ring below the membrane. Electron-dense product is seen within acidocalcisomes of cells fixed in the presence of potassium pyroantimonate which precipitates calcium. For best visualization of all acidocalcisomes, whole cells should be allowed to dry on carbon and formvar-coated grids in the transmission electron microscope, especially if it is equipped with an energy filter so that the electron spectroscopic images can be obtained (Figure 8). The first indication that the acidocalcisome is an acidic organelle, came from studies where it was shown that the inclusion vacuoles found in procyclic forms of *T. brucei* became larger when the cells were cultivated in the presence of chloroquine, an acidotropic drug. Later on it was shown by fluorescence microscopy that vacuoles of varying size found in *T. brucei* and *T. cruzi* were labeled with acridine orange and that the accumulation of this dye was sensitive to bafilomycin A, nigericin, and NH_4Cl. In the case of epimastigotes of *T. cruzi* it is important to have in mind that another acidic organelle, the reservosome, also concentrates acridine orange. The exact pH of the acidocalcisome has not yet been determined.

From a biochemical perspective, we should consider two basic components of the acidocalcisome: the matrix and the membrane. The matrix of the acidocalcisome has been mainly analysed in terms of its elemental composition, based primarily on analytical methods associated with electron microscopy. In these experiments the elemental content is compared between the inner portion of the organelle and other portions of the cell, such as the cytoplasm. The following elements have been shown to be concentrated into the acidocalcisome: P, Mg, Ca, Na and Zn, and very little Cl, K and S. The low content of S suggests a low content of proteins. Electron energy loss spectroscopy revealed the presence of P and O, suggesting the presence of carbohydrates. Recent studies have shown that the phosphorous observed in acidocalcisomes of *T. cruzi* is in the form of PPi and short-chain polyphosphates. In addition, PPi seems to be the most abundant high-energy phosphate present in *T. cruzi*. Physiological studies using permeabilized cells showed the involvement of a bafilomycin A_1-sensitive vacuolar H^+-ATPase in the process of acidification and of a vanadate-sensitive Ca^{2+}-ATPase in the uptake of Ca^{2+}. These observations were also confirmed using intact cells loaded with fura-2, a fluorescent indicator of Ca^{2+}.

The use of an immunochemical approach showed the presence of the following enzymes in the membrane of the acidocalcisomes of trypanosomatids: (a) A vacuolar H^+-ATPase. (b) A Ca^{2+}-H^+- translocating ATPase whose gene was cloned, sequenced and expressed. Antibodies generated against the protein product of the gene (Tca1) labeled the membrane of the acidocalcisome as well as the plasma membrane of *T. cruzi*. (c) A vacuolar-type proton-translocating pyrophosphatase (V-H^+-PPase) was identified and localized in the membrane of the acidocalcisome and in the plasma membrane of trypanosomatids using antibodies recognizing the enzyme found in plants. The functional properties of this enzyme have been characterized in some detail in studies analysing the effect of inorganic pyrophosphate, and pyrophospate analogues on the acidification of isolated acidocalcisomes. (d) Evidence has been obtained for the presence of a Na^+-H^+ exchanger and a Ca^{2+}-H^+ exchanger in the membrane of the acidocalcisome.

What are the functions played by the acidocalcisome? In recent reviews Docampo and Moreno (1999; 2001) considered four possibilities which are not exclusive. Indeed, based on the information available it is possible that this organelle is involved in several biological processes. These hypotheses were

(a) A role in the process of Ca^{2+} storage to be used at certain points in the parasite's life cycle. For instance, amastigote forms of *T. cruzi* live in the cytoplasm of the host cell where the concentration of Ca^{2+} is in the range of 0.1 mM which is in contrast to the trypomastigote form which lives in a environment where the concentration is around 2 mM. At this concentration it would be difficult to accumulate significant stores of Calcium. Since Ca^{2+} is involved in several necessary signaling processes (cell transformation, cell interaction, etc.), the amastigote form of *T. cruzi* may have developed a special way to accumulate this ion into acidocalcisomes, thus explaining the presence of a large number of acidocalcisomes in amastigote forms.

(b) A second role would be to act as an energy storing organelle containing a large amount of inorganic PPi. It is expected that further biochemical studies on this area will lead to new information on the bioenergetics of protozoa, opening new perspectives for the development of new parasitic drugs.

(c) In view of the presence of a H^+-ATPase in the membrane of the acidocalcisome, it can also play some role in the regulation of the cytoplasmic pH.

(d) The acidocalcisome can also play a role in the control of osmoregulation, as another organelle known as acidosome does in *Dictyostelium discoideum* - which possesses an elaborate contractile vacuole system.

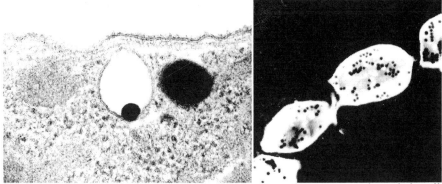

Figures 7 and 8. Acidocalcisome of T. cruzi as seen in a thin section and in whole cells, respectively (After Miranda et al., 2000).

References

Cooper R, de Jesus AR, Cross GA. 1993. Deletion of an immunodominant *Trypanosoma cruzi* surface glycoprotein disrupts flagellum-cell adhesion. J Cell Biol 122:149-56.

de Souza W. 1989. Components of the cell surface of trypanosomatids. Prog Protistol 3:87-184.

de Souza W. 1995. Structural organization of the cell surface of pathogenic protozoa. Micron 26:405-430.

Docampo R, Moreno SNJ. 1999. Acidocalcisome: a novel Ca^{2+} storage compartment in trypanosomatids and apicomplexan parasites. Parasitol Today 15:443-448.

Docampo R, and Moreno SNJ. 2001. The acidocalcisome. Mol Biochem Parasitol 33:151-159.

Englund PT, Guilbride DL, Hwa K-Y, Johnson CE, Li C, Rocco LJ, Torri AF. 1996. In *Molecular Biology of Parasitic Protozoa.* Smith DF, Parsons M. Eds. IRL Press, pp.75-87.

Flaspohler JA, Rickoll WL, Beverley SM, Parsons M. 1997. Functional identification of a *Leishmania* gene related to the peroxin 2 gene reveals a common ancestry of glycosomes and peroxisomes. Mol Cell Biol 17:1093-1101.

Gull K. 2001. The biology of kinetoplastid parasites: insights and challenges from genomics and post-genomics. Int J Parasitol 31:443-452.

Landfear SM, Ignatushchenko M. 2001. The flagellum and the flagellar pocket of trypanosomatids. Mol. Biochem. Parasitol. 115:1-17.

Miranda K, Benchimol M, Docampo R, de Souza W. 2000. The fine structure of acidocalcisomes in *Trypanosoma cruzi*. Parasitol Res 86:373-84.

Morris JC, Drew ME, Klingbeil MM, Motyka SA, Saxowsky TT, Wang Z, Englund PT. 2001. Replication of kinetoplast DNA: an update for the new millenium. Int J Parasitol 31:453-458

Opperdoes FR. 1988. Glycosomes may provide clues to the import of peroxisomal proteins. Trends Biochem Soc 13:255-260.

Parsons M, Furuya T, Pal S, Keissler P. .2001. Biogenesis and function of peroxisomes and glycosomes. Mol Biochem Parasitol 115:19-28.

Pimenta PFP, De Souza W, Souto-Padron T, Pinto da Silva P. 1989. The cell surface of *Trypanosoma cruzi*: a fracture flip, replica-staining label fracture survey. Eur J Cell Biol 50:263-271.

Porto-Carreiro I, Attias M, Miranda K, de Souza W, Cunha e Silva N. 2000. *Trypanosoma cruzi* epimastigote endocytic pathway: cargo enters the cytostome and passes through an early endosomal network before storage in reservosomes. Eur J Cell Biol 79:858-869.

Soares MJ, Souto-Padron T, de Souza W. 1992. Identification of a large pre-lysosomal compartment in pathogenic protozoan. J Cell Sci 102:157-167.

Solari AJ. 1980. Three-dimensional fine structure of the mitotic spindle in *Trypanosoma cruzi*. Chromosoma 78:239-245.

Souto-Padron T, de Souza W, Heuser JE. 1984. Quick-freeze, deep-etch rotary replication of *Trypanosoma cruzi* and *Herpetomonas megaseliae*. J Cell Sci 69:167-78.

TRYPANOSOMA CRUZI SURFACE PROTEINS

A. C. C. Frasch.
Instituto de Investigaciones Biotecnológicas-Instituto Tecnológico de Chascomús, Universidad Nacional de General San Martín and Consejo Nacional de Investigaciones Científicas y Técnicas, Casilla de Correo 30, 1650 San Martín, Provincia de Buenos Aires, Argentina

ABSTRACT

Trypanosoma cruzi expresses a number of proteins and glycoconjugates on its surface membrane. These are involved both in the protection of the parasite, and in other functions which are essential for its survival in different hosts. Some surface molecules are organised as large families and the structures of some of these family members have recently been solved.

INTRODUCTION

For cells that must actively interact with the medium, the surface membrane of the cell is more than just a barrier which separates the inside from the outside. The surface membrane is decorated with a number of macromolecules, each one with specific functions. This requirement becomes even more important for protozoan parasites which must sense and survive in generally unfavourable environments. Whether the parasite is extracellular or intracellular, it must be protected from sudden changes in the environment and simultaneously carry out all the processes required for nutrition and multiplication. The eukaryote *Trypanosoma cruzi* is no exception and its membrane is decorated by a large and diverse number of molecules. In the case of this trypanosome and other protozoan parasites, the environment radically changes when moving from one host to the other. Consequently, the composition of the surface also must change, as different functions may require different effectors.

In analyzing the complexity of a cell's surface, the first groups of molecules that come to light are the most abundant ones. In *T. cruzi*, the largest families of surface molecules have been analysed in some detail. These families may constitute a small proportion of the proteins and glycoproteins that the parasite requires for interaction with the hosts and medium, however, their characterization may shed some light on the complexity present when two different organisms live together. The complexity of this interaction is what has allowed *T. cruzi* to adapt to some defined hosts, but not to others. Thus, the components on the surface of the parasite are the consequence of the environment to which they have been exposed during evolution. The quantitatively major macromolecules on the *T. cruzi* surface are members of three large families: the mucin and mucin-like, the trans-sialidase/trans-sialidase-like proteins and the glycoinositolphospholipids (GIPLs).

MUCINS IN VERTEBRATES: STRUCTURES AND FUNCTIONS.

Mucins are glycoproteins which have a large proportion (up to 85%) of carbohydrates. They are mainly studied in vertebrate organisms and are characterized by the presence of a core protein region with a large number of short oligosaccharide side chains. The first monosaccharide is usually a N-acetylgalactosamine (GalNAc), with an O-glycosidic linkage to either a Ser or Thr residue. Consequently, mucin core proteins are rich in these amino acid residues as well as Pro and Lys residues. Areas rich in Pro-Thr-Ser- residues, PTS regions, can be found in tandem amino acid repeats or in non-repeated regions, but are usually located in the central part of the core protein. The sugar side chains typically have galactose (Gal), N-acetylgalactosamine (GalNAc), N-acetylglucosamine (GlcNAc), fucose and a terminal sialic acid. Two families (types) of mucins have been described. One family is produced by epithelial cells; these members have a large molecular mass (about 10^3-10^4 kDa), are either transmembrane or secreted, and are involved in protection of the cell. The cell is protected by the mucins located on the surface membrane or by mucins secreted into the medium which form an extracellular dense gel, due to disulfide bonds. The second mucin family was identified more recently. These members are smaller in size, are present in leukocytes and endothelial cells, and participate in lymphocyte trafficking. These are known as sialomucins since they usually have sialic acid in the non-reducing end and act as ligands for lectins. Mucins and lectins in blood cells adhere to their counterparts in endothelial cells, a step essential to allow extravasation. A number of mucins have been described in protozoan parasites such as *Leishmania spp.* and trypanosomes, with the *T. cruzi* mucins being the most extensively studied.

STRUCTURAL FEATURES OF *T. CRUZI* MUCINS.

In *T. cruzi*, mucins constitute a dense and continuous cell surface coat that is involved in protection. Different mucin types cover the surface of most developmental stages of the parasite. They are attached by a glycosylphosphatidylinositol (GPI) anchor and, thus, can be released into the medium through the action of a GPI-specific phospholipase. Two heterogeneous groups of mucins are expressed in different stages of the parasite (Figure 1).

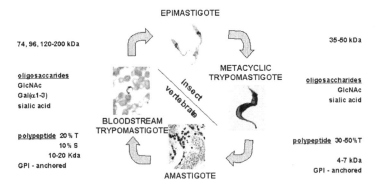

Figure 1. Mucins in the *T. cruzi* life cycle. The main characteristics of the two mucin groups, are indicated on each side of the diagram corresponding to that parasite stage. Gal: galactose, GalNAc: N-acetylgalactosamine, GlcNAc: N-acetylglucosamine, GPI-anchored: glycosylphosphatidylinositol-anchor.

One group is expressed in the insect vector stages (epimastigote and metacyclic trypomastigote) and their apparent size, ranges from 35 to 50 kDa. The second mucin group is expressed in the bloodstream trypomastigote stage, which circulates in the blood of the vertebrate host, and is composed of a large number of molecules ranging in size from 60 to 200 kDa. The carbohydrate moiety of *T. cruzi* mucins has two unusual features; a) the first monosaccharide bound to the core protein is GlcNAc instead of GalNAc, as is the case in vertebrate mucins, b) although sialic acid is frequently found in vertebrate mucins, sialic acid in *T. cruzi* is added to the membrane mucins extracellularly, and not in the Golgi apparatus by a sialyltransferase, as in vertebrate cells. The rather surprising mechanism of sialic acid incorporation to *T. cruzi* surface mucins involves a novel enzyme known as trans-sialidase, which will be described in the next section.

The large heterogeneity of both *T. cruzi* mucin groups, the 35-50 and the 60-200 kDa, is due in part to the presence of different core proteins. Mucin diversity was discovered through gene characterization, and the structure of many putative mucin core proteins was deduced. Two gene families whose members have homologies to vertebrate mucins were also identified (Figure 2).

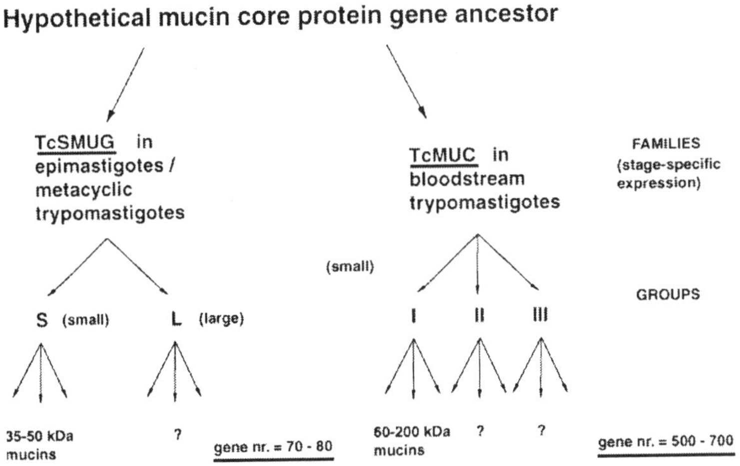

Figure 2. Hypothetical origin of the 35-50 and 60-200 kDa mucin family members. TcSMUG: *T. cruzi* small mucin genes, expressed in the vector parasite stages; TcMUC: *T. cruzi* mucin, expressed in the mammalian stages of the parasite. The interrogation marks indicate that the protein products have not been found yet.

One family is composed of about 70-80 genes, and members can be classified into two subgroups (S and L). The product of one of the subgroups (S) corresponds to the 35-50 kDa mucins expressed in the epimastigote/metacyclic trypomastigotes insect vector stages. The second gene family has about 500-700 members which can be classified into three subgroups (I, II, III). At least one of the subgroups (I) code for the 80-200 kDa mucins present in the bloodstream vertebrate form of *T. cruzi* .

While several differences are clear among these gene products they all share a similar overall structure characterized by: a) a N-terminal region that includes a signal peptide for surface location, b) a central domain rich in Thr/Ser and Pro which constitutes the target sites for O-glycosylation and c) a C-terminal region that includes the site for GPI-anchor to the surface membrane (Figure 3).

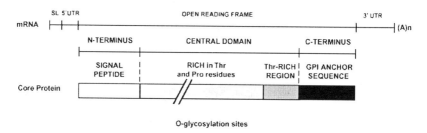

Figure 3. Overall structure of *T. cruzi* mucin mRNAs and proteins. The bar on top of the figure indicates the overall structure of the mature mucins mRNA that includes: the spliced leader (SL) sequence (added through trans-splicing to all mRNAs in trypanosomes), the 5′- and 3′- untranslated regions (UTRs) and the open reading frame coding for the N-terminus, a central domain and the C-terminus of the protein. (A)n stands for polyA+ tail in the mature mRNAs. The N-terminus includes the signal peptide for surface location and the N-terminus of the mature protein once processed. The central domain contains the region rich in threonine (Thr) and proline (Pro) residues that constitutes the O-glycosylation sites. Thr/Pro rich domain might be organized in tandems in some of the mucin groups (L and I, Figure 2).

However, there is one structural feature that differentiates the mucin core proteins in the 35-50 kDa molecules from those of the 80-200 kDa mucins: the N-terminus of the mature core protein. This N-terminus is highly conserved among the 35-50 kDa mature mucins in the epimastigotes and metacyclic trypomastigotes. Conversely, the N-terminal region is hypervariable, among the members of the 60-200 kDa mucins, expressed in the parasite found in the vertebrate host.

WHAT IS THE ROLE OF MUCINS IN *T. CRUZI*?

Mucins on the parasite surface are essentially for protection, although other functions can not be excluded (see below). The mucin coat might be considered a shield of carbohydrates covering the entire trypanosome and preventing the action of molecules in the medium. Although not proven, it is possible that flexible epitopes made of carbohydrates on the parasite surface might prevent or delay a high affinity antibody response. Furthermore, the non-reducing and exposed ends of the carbohydrate chains in mucins are decorated with sialic acid. In many pathogenic microorganisms, sialic acid is known to prevent activation of the complement pathway through the promotion of C3b cleavage and inhibition of the assembly of C3 convertase, a key component of the complement pathway. In *T. cruzi*, sialic acid appears to promote the cleavage of C3b, partially protecting the parasite from complement. In addition, the presence of sialic acid in mucins prevents the bloodstream trypomastigote from lysis by anti-alpha-galactosyl antibodies. These antibodies are produced naturally in humans and recognize terminal alpha-galactoses of mucins and lyse bloodstream trypomastigotes in infected patients. The addition of sialic acid to terminal beta-galactoses, also present in the parasite mucins, prevents the binding of anti-alpha-galactosyl antibodies and the lysis of the parasite.

The sialic acid moiety of the trypomastigote mucins was suggested to be involved in the invasion of mammalian host cells by the parasite. Both the presence of the sugar or its removal from surface mucins, the latter process exposing terminal galactoses, were found to increase the binding to and/or invasion into mammalian cells. Thus, it is likely that mucins can act both as anti-adhesive molecules and as adhesive molecules, as is the case in other cells. Furthermore, mucins and/or their GPI-anchor moiety might function in the adherence of the parasite to macrophages and in the production of nitric oxide and cytokines by these cells.

In addition to exposing carbohydrates on the parasite surface, mucins also expose the N-terminal end of the core protein. This part of the protein is antigenic. Interestingly, the mucins of the parasite form in the mammalian host, the 60-200 kDa molecules, have N-terminal regions that are heterogeneous among molecules. It has been suggested that this hypervariability in the N-terminus of mucins is required to prevent an efficient immune response against the parasite. In accordance with this idea, the N-terminus of mucins expressed by the parasite forms in the insect vector, the 35-50 kDa molecules, is highly homogenous.

A number of the functions of surface mucins in *T. cruzi* is due to the presence of the terminal sialic acid in the carbohydrate moieties of these glycoproteins. Sialic acid is incorporated by the action of an extracellular surface enzyme called trans-sialidase. This enzyme is a modified sialidase that is part of the second large family of molecules located on the parasite surface.

SURFACE SIALIDASES IN MICROORGANISMS.

Sialidases are hydrolytic enzymes which release sialic acid from sialoglycoconjugates. A number of microorganisms have these sialidases on their extracellular surface. Cell surface sialidases have been identified in viruses (influenza), pathogenic bacteria (*Vibrio cholerae*, *Clostridium perfringens* and *Salmonella typhimurium*), protozoan parasites (trypanosomes) *Tritrichomonas*, and in metazoa (the leech *Macrobdella decora*). In most cases, sialidases are used for nutrition but in others the surface sialidase is involved in pathology. As in the case of the influenza enzyme it releases the virus from the surface of the infected cell, spreading the infection. In intracellular bacterial infections, surface sialidases are also required for unmasking binding sites allowing cell invasion.

Sialic acid is a generic term used to designate a large family of the nine-carbon sugar N-acetyl neuraminic acid. Sialic acid is usually linked to a galactose through alpha-2,3 or -2,6 bonds. Some sialidases are able to hydrolyze both linkages whereas others prefer alpha-2,3 substrates like the influenza and *S. typhimurium* enzymes. There are some sialidases that act only on alpha-2,3 sialic acid like the ones from leech and trypanosomes. Sialidases from *T. cruzi* and *T. brucei* (the agent of sleeping sickness in Africa) have a somewhat different enzymatic activity. Instead of releasing sialic acid, they transfer the monosaccharide to a terminal galactose in a glycoconjugate. Because they are transferases rather than hydrolases they were named trans-sialidases. They are not sialyltransferases, which are enzymes located in the Golgi apparatus in eukaryotic cells, because trans-sialidases cannot use CMP-sialic acid, the only donor of sialic acid in the sialyltransferase reaction. Trans-sialidases are highly specific for α-2,3 sialic acid and transfer the sugar to terminal beta-galactose acceptors. The linkage of the product is also α-2,3, and the reaction is fully reversible. Donors of the sialic acid are glycoconjugates in the medium (blood of the vertebrate host) and the acceptors are the parasite mucins located on the surface membrane.

Viral, bacterial and trypanosomal sialidases have the same overall tertiary structure, but are not highly conserved in primary structure. Bacterial and trypanosome sialidases share about 30% of identity at the amino acid level and contain two conserved sequence motifs. The first is the FRIP (Phe-Arg-Ile-Pro) motif and the second is the asparagine box (Ser/Thr-X-Asp-X-Gly-X-Thr-Trp), which can be present in 1 to 5 copies per molecule. The tertiary structure of sialidases is conserved among enzymes from different microorganisms. The monomer has a superbarrel or beta-propeller structure. It is made from six four-stranded and antiparallel beta sheets arranged around a central axis that looks like the blades of a propeller. Sialidases are about 350-400 amino acid residues long (about 40kDa) but some enzymes have extra-domains explaining their larger size (about 100 kDa). Examples of "large" sialidases are those present in *V. cholerae*, *M. decora* and trypanosomes. This extra domain has in general the structure of plant lectins. Their function is unknown.

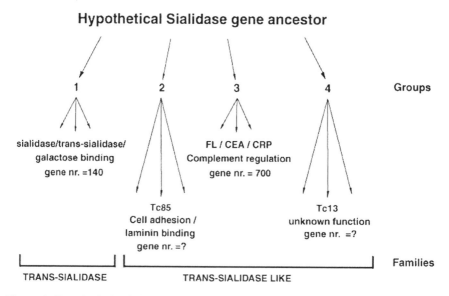

Figure 4. Hypothetical origin of the trans-sialidase and trans-sialidase like families. The two families are indicated with bars on the bottom of the figure. The four groups of genes/proteins are indicated with numbers, together with the function described for some members of each group/family and the gene number if known.

THE *T. CRUZI* TRANS-SIALIDASE AND TRANS-SIALIDASE-LIKE FAMILY.

The second large surface protein family is composed of molecules having sequence homologies with sialidases. About 1000 genes having this overall structure are present on the parasite genome. The products of these genes analyzed so far are anchored to the surface membrane by a GPI structure, and are released into the medium. Some of these proteins have sialidase or, mainly, trans-sialidase activity (see above). However, other members of this family have diverged enough as to have acquired functions unrelated to these enzymatic activities.

This family of surface proteins was also characterized through genetic analysis. As in the case for mucins, members of the trans-sialidase and trans-sialidase-like family all share sequence homologies. Proteins which have more than 95% sequence identity to those that display sialidase/trans-sialidase activity (Figure 4) will be included in the trans-sialidase family.

They constitute a group of about 140 proteins. The second group of proteins, the trans-sialidase like family, has about 30 % of identity with trans-sialidases and is organized into three groups according to their sequence homologies (Figure 4). They lack enzymatic activity but have other functions, as described below. Thus, a number of proteins having radically different functions might have originated from a hypothetical sialidase gene ancestor (Figure 4).

All these proteins have the motifs conserved among bacterial sialidases; namely the FRIP (Phe-Arg-Ile-Pro) motif and 1 to 4 copies of the Asp (Ser-X-Asp-X-Gly-X-Thr-Trp) box (Figure 5). In addition, they all share another motif specific to the trypanosomal sialidases: Val-Thr-Val-X-Asn-Val-X-Leu-Tyr-Asn-Arg (Figure 5). The function of this highly conserved motif is unknown. Some members within the four groups have extra domains with amino acid repeats units in tandem (Figure 5).

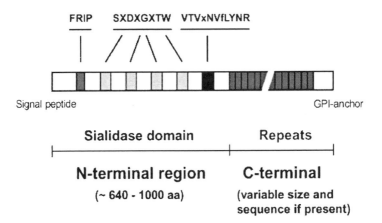

Figure 5. Overall structure of members of the trans-sialidase and trans-sialidase like families. A picture of the structure of proteins from both families is shown. The three conserved motifs are indicated on the top of the figure (see text). The sialidase domain indicates the N-termini of proteins having regions with homologies to bacterial sialidases, including some of the conserved motifs. The C-terminal domain, possessing amino acid repeat units in tandem, is present in some, but not all, members of both families.

One of the best characterized is the repeated domain present in the trans-sialidase itself (group I) which was named SAPA for shed acute phase antigen. This domain is highly immunogenic and is involved in keeping the enzyme active in the blood of the host. The 3-dimensional structure of a trypanosomal (*T. rangeli*) sialidase and the modelled structure of *T. cruzi* trans-sialidase were recently resolved. Both have the beta-propeller structure typical of viral and bacterial sialidases (first 380 amino acid residues on the

N-terminus). They also have an extra domain with a lectin like structure, as is the case for the "large" bacterial sialidases.

MEMBERS OF THE TRANS-SIALIDASE AND TRANS-SIALIDASE LIKE FAMILY HAVE DIVERSE FUNCTIONS.

Although all members in these families might have originated from a hypothetical sialidase gene ancestor, they have diverged largely enough as to have acquired a number of other functions (Figure 4). Among the proteins in the trans-sialidase family, some of them have a natural single point mutation (Tyr342-His) that renders them enzymatically inactive. These inactive proteins are able to bind terminal galactoses, suggesting that they might act as lectins on the parasite surface. Members of the trans-sialidase-like family (group Tc85) are also binders of laminin or are involved in cell-cell interactions (Figure 4). Members of another group (Fl/CEA/CRP) have an apparently different function; they are involved in complement regulation. Three proteins having complement regulatory activity were identified by their ability to inhibit the assembly and/or accelerate the decay of C3 convertase, the central enzyme in the complement cascade leading to membrane pores and cell lysis. These proteins expressed on the surface of trypomastigotes might be involved in preventing both the action of the alternative and the classical complement pathways. Although complement regulation might be seen as a function unrelated to trans-sialidase activity, it is involved in protection of the parasite against the complement pathway. Through the action of trans-sialidase, the presence of sialic acid on the parasite surface, also regulates activation of the alternative complement pathway. In sum, all members of this large family have so far found to be binders of sugars or other macromolecules in the medium or in the host-cell surface or to prevent the action of complement. Consequently, they are all involved in cell invasion and/or in protection of the parasite and, in this context, might be considered virulence factors.

These proteins have to be exposed on the parasite surface and, as a consequence, are targets of the immune response of the host. In the case of mucins, diversity of the exposed protein epitopes might be a way to prevent an effective antibody response, and thus might constitute a mechanism of immune evasion. Sequence divergence among proteins of the trans-sialidase and trans-sialidase like families might also be involved in the prevention of an effective response against the parasite. They stimulate a polarized Th1 response. However, the presence of variant T cell epitopes might end in anergy, further contributing to the survival of the trypanosome.

GLYCOSYLPHOSPHATIDILINOSITOLS (GPIs) AND GLYCOINOSITOLPHOSPHOLIPIDS (GIPLs)

Mucin and trans-sialidase members are all anchored by GPI to the cell surface. In fact, most protozoa utilize GPI anchors more than other forms of membrane attachment. The common structural motif in GPIs from trypanosomes and leishmanias are Man alpha 1-4GlcN alpha 1-6-myo-inositol-1-HPO4-lipid, defining the members of what was suggested to be a GPI superfamily. In *T. cruzi*, as well as in other protozoa, the structure of GPIs varies according to the developmental stage of the parasite. For example, while ceramide is the only constituent in some parasite stages, others have ceramide and alkylglycerol. Other differences are the presence of additional galactose residues and unsaturated, instead of saturated, fatty acids in the alkylacylglycerolipid component. Differences in structures explain why GPIs from some parasite stages are potent proinflammatory agents while others lack this activity.

Glycoinositolphospholipids (GIPLs) are, together with mucins, the two most abundant glycoconjugates present in all *T. cruzi* developmental stages. GIPLs were identified 25 years ago and were formerly known as lipopeptidophosphoglycans. They all have a similar overall structure with Man4GlcN-Ins-PO$_4$ (four mannoses followed by glucosamine, inositol and a phosphate group that links this structure to the lipids in the membrane). GIPLs were found to be immunomodulatory molecules activating B-lymphocytes and natural killer cells, activities that might play an important role during *T. cruzi* infection.

OTHER SURFACE MACROMOLECULES IN *T. CRUZI*.

A number of other surface molecules have been described in *T. cruzi*; most of them might be present in smaller amounts. This does not say anything about their relative importance in the parasite life cycle. In fact, a few of them were suggested to be highly relevant in the host-parasite interaction. A membrane associated proline racemase in the infective parasite stage was found to be a B-cell mitogen involved in polyclonal activation leading to immune evasion. Other surface proteins have not yet been assigned a function but are stage-specific, like amastin for the intracellular amastigote stage. Yet, one of the first surprises when analyzing antigens from *T. cruzi* was the finding that most of them are made up of tandem amino acid repeats. Some of these repetitive proteins are located on the parasite surface, similar to what was found in *Plasmodium* spp. The functions of these repetitive proteins are difficult to discern since they are parasite specific, lacking any homology with proteins in other cells. As is the case of the SAPA repeats and the trans-sialidase, it is possible that repetitive domains are linked to other domains having a defined function. In any case, these highly repetitive surface and internal antigens were of great utility in designing tests to diagnose *T. cruzi* infection.

SURFACE PROTEINS FROM OTHER AMERICAN TRYPANOSOMES.

Almost no information is available on other trypanosomes in America with the exception of *T. rangeli*, a trypanosome not pathogenic to humans. One large family of surface proteins was identified in this trypanosome and, interestingly, all analysed members share sequence identities with bacterial sialidases. Some of their members display sialidase activity but, unlike those molecules in *T. cruzi*, they are completely devoid of trans-sialidase activity. Other members lack sialidase/trans-sialidase activity and their function is unknown. The 3-dimensional structure of the sialidase from *T. rangeli* was the first one resolved in trypanosomes. Since its primary structure is 70% identical to that of *T. cruzi* trans-sialidase, comparison of both enzymes has provided information on why trans-sialidase is such an efficient enzyme in the transferring of sialic acid.

REGULATION OF THE EXPRESSION OF SURFACE PROTEINS IN *T. CRUZI.*

Little is known on the regulation of protein expression in *T. cruzi* and other trypanosomes. This is a subject of great interest in trypanosomatids. Unlike higher eukaryotic cells that regulate protein expression at the level of transcription initiation, trypanosomes seem to use post-transcriptional mechanisms for this purpose. Among them, regulation of messenger RNA stability seems to be the most relevant one.

Mucin transcripts expressed in the epimastigote stages (35-50 kDa mucins, group S in Figure 2), have sequences on their 3′-untranscribed regions (3′UTRs) that either stabilize or destabilize the mRNAs. The presence of certain sequences increases the half-life of transcripts, increasing the possibility of translation. Conversely, other sequences decrease the half-life of the transcripts (mRNA degradation), preventing their translation into proteins. If these sequences are present in the same mucin mRNA, how is regulation of protein expression achieved? The answer depends on the parasite stage studying question. Different stages express different RNA binding proteins that, through recognition of the sequences in the transcript, either stabilize or destabilize it. For example, there is an AU-rich sequence (named ARE for AU rich element) in the 3′UTR from S mucin transcript that destabilizes the transcript in the trypomastigote stage but not in the epimastigote stage. Destabilization is likely to be due to the expression of an RNA binding protein in trypomastigotes that recognizes this sequence and promotes its degradation. This RNA binding protein is not expressed in the epimastigote stage, rendering it more stable and allowing translation into a mucin core protein. This is an oversimplification of the situation as there might be a large number of RNA binding proteins interacting with a given transcript. Molecules involved in the modification of the half-life of mRNA have been studied for some transcripts in higher eukaryotic cells, such as protooncogenes and cytokines. Regulation of expression of these proteins are also achieved through mRNA stabilization-destabilization.

FINAL REMARKS

Protozoan parasites sense and interact with the medium through surface and/or shed-secreted molecules. If we learn which are the essential molecules involved in these functions, we might be in a better position to identify targets for the design of new chemotherapies and immune interventions. This is the practical view on to the rationale behind the further continuation of these lines of research. But there are other reasons. Among them is the better understanding of molecules and mechanisms selected during evolution to allow this parasite to survive in their hosts. Similar mechanisms might have been used also in the relationship between cells in complex organisms like ours. Mucins and lectins are used in vertebrates to allow interaction between blood and endothelial cells in the process of extravasation from blood vessels. *T. cruzi* might use mucins and lectin-like molecules to interact with vertebrate cells. In another example, the assembly of C3 convertase is under tight control by proteins such as decay-accelerating factor (DAF) in vertebrate cells to prevent damage of autologous cells. *T. cruzi* expresses a protein having 27% similarity with human DAF (a member of the trans-sialidase like family) that is also involved in accelerating the decay of both the alternative and classical complement pathway of C3 convertase. But this might be the minimal part of all processes common in the interrelation between parasite-host cells and between our own cells.

Acknowledgements: I thank Dr. J.J. Cazzulo and G. Gotz for careful reading of this chapter and P. Briones for his help in designing of the figures.

Further Reading

Almeida IC, Gazzinelli R, Ferguson MA, Travassos LR. 1999. *Trypanosoma cruzi* mucins: potential functions of a complex structure. Mem Inst Oswaldo Cruz. 94 Suppl 1:173-6.

Campetella O, Sanchez DO, Cazzulo JJ, Frasch ACC. 1992. A superfamily of *Trypanosoma cruzi* surface antigens. Parasitol Today 8:378-81.

Colli W, Alves MJ. 1999. Relevant glycoconjugates on the surface of *Trypanosoma cruzi*. Mem Inst Oswaldo Cruz 94 Suppl 1:37-49.

Ferguson MA. 1999. The structure, biosynthesis and functions of glycosylphosphatidylinositol anchors, and the contributions of trypanosome research. J Cell Sci 112:2799-809.

Frasch ACC. 2000. Functional diversity in members of the trans-sialidase and mucin families in *Trypanosoma cruzi*. Parasitol Today 16:282-86.

Fukuda M, Tsuboi S. 1999. Mucin-type O-glycans and leukosialin. Biochim Biophys Acta 1455:205-17.

Hicks SJ, Theodoropoulos G, Carrington SD, Corfield AP. 2000. The role of mucins in host-parasite interactions. Part I - protozoan parasites. Parasitol Today 16:476-81.

Krautz GM, Kissinger JC, Krettli AU. 2000. The targets of the lytic antibody response against *Trypanosoma cruzi*. Parasitol Today 16:31-4.

Pereira-Chioccola VL, Schenkman S. 1999. Biological role of *Trypanosoma cruzi* trans-sialidase. Biochem Soc Trans 27:516-8.

Perez-Vilar J, Hill RL. 1999. The structure and assembly of secreted mucins. J Biol Chem 274:31751-4

Schenkman S, Eichinger D, Pereira ME, Nussenzweig V. 1994. Structural and functional properties of Trypanosoma trans-sialidase. Annu Rev Microbiol 48:499-523.

Taylor G. 1996. Sialidases: structures, biological significance and therapeutic potential. Curr Opin Struct Biol 6:830-7 .

Tomlinson S, Raper J. 1998. Natural human immunity to trypanosomes. Parasitol Today 14:354-59.

SIGNALING IN *TRYPANOSOMA CRUZI*

R. Docampo and S. N. J. Moreno
Laboratory of Molecular Parasitology, Department of Pathobiology, University of Illinois at Urbana-Champaign, 2001 South Lincoln Avenue, Urbana, IL 61802, USA

ABSTRACT

Many signaling pathways known to be present in metazoans are also present in unicellular organisms, and *Trypanosoma cruzi* is not an exception. *T. cruzi* shares key molecules and biochemical pathways with higher eukaryotes. For example, calcium, adenylyl cyclase, guanylyl cyclase, phospholipase C, nitric oxide synthase, protein kinases, and protein serine/threonine and tyrosine phosphatases, were all found to play central roles in *T. cruzi*. However, current evidence suggest that *T. cruzi* might lack some molecules involved in signaling such as receptor-linked phospholipases, protein kinases and phosphatases, heterotrimeric G proteins, and transcription factors necessary for regulation of gene expression.

INTRODUCTION

Unicellular organisms continually sense their surrounding environment and make changes on the basis of that information. *Trypanosoma cruzi* is not an exception since it must be able to sense which nutrients are nearby and regulate its metabolic processes accordingly. It also needs to sense the proximity of cells, to be able to attach and invade them; and to sense the extracellular matrix, to be able to migrate through surrounding tissues to infect adjacent cells, or reach the bloodstream and home towards specific organs. During its life cycle, *T. cruzi* has to adapt to environments of different temperature, osmolarity, ionic composition, and pH, and some of these adaptation processes are paralleled by morphological and functional changes. These processes all require the transfer of information from detection systems through intermediate molecules within the cell (second messengers), to cause changes in the expression of genes and the activity of enzymes. In addition, as with other unicellular eukaryotes, when exposed to certain external changes such as shifts in temperature or exposure to human complement, *T. cruzi* is able to undergo programmed cell death or apoptosis. Although critical to growth, differentiation, death, and metabolic responses, very little is known about the signaling mechanisms involved in these processes in *T. cruzi*. In multicellular and some unicellular eukaryotes, signaling pathways culminate in the activation of transcription factors. In *T. cruzi*, no such transcription factors have been found and it is not known how signaling results in regulation of gene expression.

RECEPTORS

Signaling pathways start with receptor proteins, commonly found at the cell surface, that are able to bind a ligand or sense a change in the environment and transfer a signal across the plasma membrane. Two putative ligands which result in increased cyclic AMP (cAMP) levels and enhance differentiation and one possible target gene whose expression is stimulated by cAMP have been identified in *T. cruzi* (reviewed in Flawiá et al., 1997; Taylor et al., 1999; Parsons and Ruben, 2000). One of the putative ligands is a globin derived factor, a small peptide which results from proteolytic cleavage of chicken α^D-globin within the hindgut of the insect vector. Addition of this peptide results in activation of the parasite adenylyl cyclase in membrane fractions and enhanced differentiation from the epimastigote to the metacyclic trypomastigote stage. Similarly, when metacyclic trypomastigotes are exposed to fibronectin they proteolytically degrade it into several peptides, two of which appear to activate adenylyl cyclase and initiate their differentiation to the amastigote stage. The putative target gene is a cAMP inducible gene expressed in trypomastigotes but not in epimastigotes.

These studies suggested that *T. cruzi* adenylyl cyclases might share common regulatory mechanisms with the mammalian enzymes including activation by heterotrimeric G-proteins. However, when *T. cruzi* adenylyl cyclases were identified at the molecular level it was found that their predicted structure consisted of a large *N*-terminal domain, which is presumed to be extracellular, followed by a single membrane-spanning helix and an intracellular catalytic domain (Taylor et al., 1999). This is in contrast to the typical 12-transmembrane spanning structure of G-protein coupled adenylyl cyclases. This structure also suggested that they might function as catalytic receptors. The probability that the extracellular receptor domain can directly activate the intracellular cyclase component implies that there is no necessity for activation by heterotrimeric G-proteins or other regulatory factors. Thus far, heterotrimeric G-proteins have not been directly identified in *T. cruzi*, and only indirect circumstantial evidence for their presence has been presented (Flawiá et al., 1997). Adenylyl cyclases, like many surface molecules in *T. cruzi*, are encoded by a complex multigene family (Taylor et al., 1999). The *T. cruzi* adenylyl cyclase catalytic domain is constitutively active, however, in the absence of eukaryotic regulatory factors, which implies that *in vivo* these enzymes are maintained in an inactive state by a negative regulator. In the case of similar enzymes from *T. brucei*, this regulator has been postulated to be pyrophosphate (PP$_i$) (Bieger and Essen, 2001). If this were the case, *T. cruzi* adenylyl cyclase activity would be solely under metabolic control and would not require external stimulation.

Other studies (reviewed in Parsons and Ruben, 2000) have suggested that *T. cruzi* expresses receptors for alpha and beta-adrenergic agonists and that binding of specific agonists to these receptors modifies the infective capacity of the parasite *in vitro* and its cyclic AMP levels. The presence of nicotinic acetylcholine receptors was also postulated on the basis of (1) binding of cholinergic agonists, such as carbamoyl choline (carbachol), which resulted in changes in polyphosphoinositides and phosphatidic acid as well as variations of inositol phosphates apparently mediated by phospholipase C activation. (2) binding of the cholinergic antagonist nicotine, which was shown to cause changes in calcium concentration in epimastigotes (Bollo et al., 2001). The chicken α^D-globin fragment also stimulated the formation of inositol 1,4,5-trisphosphate. However, the mechanisms involved in the transduction of these signals have not been elucidated and the receptors have not been identified at the molecular level.

An L-glutamate channel receptor subtype specific for *N*-methyl-D-aspartate (NMDA), possibly involved in Ca^{2+} influx, was also postulated to be present in epimastigotes (reviewed in Flawiá et al., 1997) on the basis of: (1) an stimulatory effect of L-glutamate and NMDA on the conversion of L-[^{3}H]arginine to [^{3}H]citrulline and generation of nitrite (a derivative of NO), its inhibition by NMDA antagonists such as MK-801 and ketamine, and its stimulation by glycine, a potentiator of L-glutamine responses; (2) the stimulatory effect of L-glutamate and NMDA on soluble guanylyl cyclase activity; and (3) the specific binding of [^{3}H]MK-801 to epimastigote cells and membranes that could be displaced by unlabeled MK-801 or ketamine. This binding was also enhanced by glycine and L-serine. The stimulatory effect of L-glutamate on nitrite formation and its inhibition by MK-801 was recently confirmed by other authors (Piacenza et al., 2001). The receptor, however, has not yet been identified at the molecular level and it is not known whether its stimulation leads to intracellular Ca^{2+} increase in the parasite.

　　T. cruzi also possesses molecules involved in response to environmental stresses, that do not necessarily depend on surface receptors. These include heat-shock proteins, and molecules involved in cell cycle regulation such as cyclin-dependent and other cycle regulated kinases, cyclins, and other regulatory proteins (reviewed in Parsons and Ruben, 2000).

INTRACELLULAR MEDIATORS

　　Receptors pass on the signals to other molecules. As we mentioned above, except for receptor-type adenylyl cyclases, other molecules that transduce signals following interactions with specific extracellular ligands have not yet been found in *T. cruzi*. A phosphatidyl inositol phospholipase C (PI-PLC) has been identified at the molecular level (Furuya et al., 2000). However, as occurs with other PI-PLC of early branching eukaryotes, it is of the δ-type, probably regulated only by Ca^{2+} or small G proteins. There is also no evidence of receptor-type protein kinases or phosphatases in *T. cruzi*. Two serine/threonine protein phosphatases have been identified in *T. cruzi* (Orr et al., 2000). One of them, homologous to mammalian PP1, was expressed and shown to be inhibited by low concentrations of calyculin A (IC_{50}, ≈ 2 nM). Low concentrations of calyculin A (1-10 nM) caused growth arrest of epimastigotes, that were shown to be incapable of undergoing cytokinesis (Orr et al., 2000) and stimulated differentiation of trypomastigotes to amastigotes (Grellier et al., 1999). It is not known how the activity of these phosphatases is regulated and which are their effectors. Protein tyrosine phosphatase activities have also been identified in *T. cruzi* (reviewed in Furuya et al., 1998), but they were not characterized at the molecular level. One of these activities is located in the external surface of infective stages of *T. cruzi* and possibly has a role in invasion of mammalian cells, since invasion results in tyrosine dephosphorylation of host proteins and is inhibited by inhibitors of this protein tyrosine phosphatase (Zhong et al., 1998). Finally, a gene with homology to a protein kinase B, encoding an enzyme highly specific for threonine phosphorylation and probably lipid anchored to the membrane of different stages of *T. cruzi*, was also found. Its role, as in other organisms, is currently unknown (Pascuccelli et al., 1999). It is interesting to note that a phosphatidylinositol 3-kinase (PI3K), an enzyme that generates phospholipids that activate protein kinase B in other organisms, is apparently present in *T. cruzi* trypomastigotes since treatment with PI3K inhibitors prior to infection reduced parasite entry into culture cells (Wilkowsky et al., 2001).

　　Interactions between signaling proteins through specific recognition domains, allowing the formation of protein complexes and resulting in changes in the localization or activity of enzymes have not been described in *T. cruzi*. No proteins have yet been found containing domains such as the SRC homology regions 2 and 3 (SH2 and SH3 domains, respectively) or pleckstrin homology (PH) domains. Nor is there evidence for the so-called 'adaptor' or 'docking' proteins that bring together other signaling molecules.

Cyclic AMP

Cyclic AMP (reviewed in Flawiá et al., 1997; Parsons and Ruben, 2000) has been found to be important for the differentiation of epimastigotes to metacyclic trypomastigotes and from trypomastigotes to amastigotes. A calmodulin-dependent phosphodiesterase, which returns cyclic AMP to basal levels and a protein kinase A, which is the effector of cyclic AMP, were identified in *T. cruzi*. The targets of protein kinase A have not yet been identified and molecular characterization of these proteins has not yet been accomplished. A cyclic AMP regulated gene was identified but the function of the encoded protein is unknown.

Nitric oxide

A nitric oxide synthase (NOS) (reviewed in Flawiá et al., 1997) was partially purified from soluble extracts of epimastigotes and shown to require NADPH and to be stimulated by Ca^{2+}, calmodulin, tetrahydrobiopterin and FAD; and inhibited by N--methyl-L-arginine (L-NAME). A sodium nitroprusside-activated guanylyl cyclase activity was also detected in cell-free, soluble preparations of epimastigotes. It was postulated that Ca^{2+} entry through the putative L-glutamate receptor would stimulate NOS generating NO which would stimulate the soluble guanylyl cyclase leading to an increase in cyclic GMP levels. The effectors of cyclic GMP (cyclic GMP dependent protein kinases) have not been identified. NO generation has been linked to flagellar motility in epimastigotes. Recently, NO was also postulated to mediate protection of epimastigotes against apoptotic death induced by 10% fresh human serum although the mechanism for this effect was not elucidated (Piacenza et al., 2001).

Inositol phosphate/diacylglycerol pathway

Phosphoinositide-specific phospholipases C (PI-PLCs) catalyze the hydrolysis of phosphatidylinositol-4,5-bisphosphate (PIP_2) to D-*myo*-inositol-1,4,5-trisphosphate (IP_3) and *sn*-1,2-diacylglycerol (DAG). Both products of this reaction function as second messengers in eukaryotic signal transduction cascades. The soluble IP_3 triggers release of calcium from intracellular stores. The membrane-resident DAG controls cellular protein phosphorylation states by activating various protein kinase C isozymes. Four classes of mammalian PI-PLCs with 11 different isozymes have been characterized ($\beta1$-$\beta4$; $\delta1$-$\delta2$, $\gamma1$-$\gamma4$, and ϵ). The activity of β- and γ- isozymes (MW 145-150 kDa) is regulated by G protein-coupled and tyrosine kinase-linked receptors, respectively. The activity of ϵ-PI-PLC (MW 255 kDa) is apparently regulated by Ras (Kelley et al., 2001). These isozymes are related to the much smaller δ-isozymes (MW ~85 kDa). It seems very likely that PI-PLC-δ evolved first, because every PI-PLC cloned so far from a non-mammalian species (for example *Dictyostelium, Chlamydomonas,* yeast, higher plants) is clearly a δ-isoform. It is currently not known how δ-isozymes are regulated *in vivo*. It is possible that they are regulated only by calcium ions although some PI-PLC-δ also appear to be regulated by small GTP-binding proteins (reviewed in Furuya et al., 2000).

The presence and operation of the inositol phosphate/diacylglycerol signaling pathway (reviewed in Furuya et al., 2000) was demonstrated in epimastigotes of *T. cruzi*. IP3 and DAG formation was stimulated by Ca^{2+} in digitonin-permeabilized cells, thus suggesting the presence of a PI-PLC. The presence of different inositol phosphates in amastigotes and trypomastigotes was reported later. A shift in the levels of phosphoinositide metabolites after incubation of epimastigotes with carbachol and the stimulation of IP3 and DAG production and epimastigote proliferation by fetal calf serum were also reported. This last effect was enhanced by AlF_4^-, GTP, or non-hydrolyzable analogs of GTP, suggesting G protein mediation of this phenomenon. A PI-PLC activity was detected in epimastigote lysates using phosphatidylinositol (PI) as substrate. A phosphoinositide-specific phospholipase C gene (*TcPI-PLC*) was cloned, sequenced, expressed in *E. coli*, and the protein product (TcPI-PLC) was shown to have structural and enzymatic characteristics similar to those of plant δ-type PI-PLCs. The *TcPI-PLC* gene is expressed at high levels in the epimastigote and amastigote stages of the parasite, and its expression is induced during the differentiation of trypomastigotes into amastigotes, where it associates to the plasma membrane and increases its catalytic activity. In contrast to other PI-PLCs described so far, the deduced amino acid sequence of TcPI-PLC revealed some unique features such as an *N*-myristoylation consensus sequence at its *N*-terminal end, lack of an apparent pleckstrin homology (PH) domain and a highly charged linker region between the catalytic X and Y domains. TcPI-PLC is lipid modified *in vivo* as demonstrated by metabolic labeling with [³H]myristate and [³H]palmitate and fatty acid analysis of the immunoprecipitated protein, and may constitute the first example of a new group of PI-PLCs. It was hypothesized that the enzyme is involved in the signaling cascades that take place during the differentiation processes of *T. cruzi*.

A protein kinase C was characterized in *T. cruzi* epimastigotes (reviewed in Flawiá et al., 1997). This enzyme requires phosphatidylserine and Ca^{2+} for activity and is stimulated by DAG and is apparently of the conventional (α) protein kinase C family (Gomez et al., 1999). The *T. cruzi* protein kinase C has not been characterized at the molecular level and the proteins phosphorylated by protein kinase C have not been identified. Although DAG could stimulate a protein kinase C activity in *T. cruzi*, IP₃ was unable to release Ca^{2+} from intracellular stores of the parasite (reviewed in Parsons and Ruben, 2000).

Calcium
There is compelling evidence that calcium ions play a crucial role in controlling many biological processes. The free cytosolic Ca^{2+} concentration is the key variable governing the intracellular actions of Ca^{2+}. As in most eukaryotic cells Ca^{2+} homeostasis in *T. cruzi* is achieved by the concerted operation of several Ca^{2+}-transporting systems located in the plasma membrane, endoplasmic reticulum and mitochondria. In addition, most of the intracellular Ca^{2+} in *T. cruzi* is located in an acidic organelle which has been named the acidocalcisome and which also possesses Ca^{2+}-transporting mechanisms (reviewed in Docampo and Moreno, 2001).

In the plasma membrane there is a specific Ca^{2+}-ATPase (PMCA-type) which functions in the active export of the ion. The gene encoding this protein has been described and the enzyme is also localized in the acidocalcisome. However, little is known about the mechanism of Ca^{2+} entry, besides the description of a putative L-glutamate receptor that could act as a Ca^{2+} channel. Ca^{2+} entry has been shown to be stimulated by fatty acids

(Catisti et al. 2000) but there is no firm evidence for the presence of a "capacitative" or store-operated Ca^{2+} uptake. The inner mitochondrial membrane possesses a uniport carrier for calcium, which allows the electrogenic entry of the cation driven by the electrochemical gradient generated by respiration or ATP hydrolysis. Calcium efflux, on the other hand, takes place by a different pathway that appears to catalyze the electroneutral exchange of internal calcium by external protons or sodium.

In the endoplasmic reticulum the influx is catalyzed by a SERCA-type Ca^{2+}-ATPase whose molecular characterization has recently been accomplished (Furuya et al., 2001). The pathway for calcium release from the endoplasmic reticulum is still poorly characterized since calcium release from the endoplasmic reticulum was not observed in the presence of IP3 or several other second messengers tested. Ca^{2+} uptake into the acidocalcisomes is catalyzed by the PMCA-type ATPase, also localized in the plasma membrane. It is not known how Ca^{2+} is released from acidocalcisomes. These organelles are very rich in short and long chain polyphosphate (polyP) and it has been shown that polyP hydrolysis occurs in association with Ca^{2+} release and this mechanism could be involved in controlling intracellular Ca^{2+} (Ruiz et al., 2001). Many of the intracellular actions of calcium occur by binding to low molecular weight calcium-binding proteins. The calcium-binding protein calmodulin appears to be particularly important and it has been found in *T. cruzi*. *T. cruzi* also possesses a calmodulin-dependent protein kinase activity similar to that of brain CaM KII (reviewed in Flawiá et al., 1997).

Ca^{2+} signaling has been shown to play a key role in the process of mammalian cell invasion and the intracellular development of *T. cruzi*. An increase in the cytosolic Ca^{2+} concentration ($[Ca^{2+}]_i$) of *T. cruzi* trypomastigotes occurs upon invasion, and pretreatment of the trypomastigotes with intracellular Ca^{2+} chelators (BAPTA/AM or Quin 2/AM) to prevent the increase in $[Ca^{2+}]_i$ results in an inhibition of cellular invasion (reviewed in Docampo and Moreno, 1996).

CONCLUSIONS

Table 1 provides a summary of the main signaling molecules and pathways known to be present in *T. cruzi* and their possible roles. It is expected that completion of the *T. cruzi* genome project coupled to its functional analysis will increase our knowledge of signaling mechanisms in this important parasite.

Table 1. Signaling pathways present in *Trypanosoma cruzi*

Enzyme/process	Second messenger	Effector	Function
Adenylyl cyclase	Cyclic AMP	Protein kinase A	Differentiation
Guanylyl cyclase	Cyclic GMP	Protein kinase	Unknown
Phospholipase C	IP_3, DAG	Calcium stores/	Invasion/
		Protein kinase C	Differentiation
Nitric oxide Synthase	NO	Guanylyl cyclase	Anti-apoptosis/
			Flagellar motility
PI-3-kinase	Phospholipids	Protein kinase B	Invasion
Ca^{2+} entry/release	Ca^{2+}	CaM/CaM kinases	Invasion

References

Bieger B, Essen L-O. 2001. Structural analysis of adenylate cyclases from *Trypanosoma brucei* in their monomeric state. EMBO J 20:433-45.

Bollo M, Venera G, Bonino MB de J, Machado-Domenech E. 2001. Binding of nicotinic ligands to and nicotine-induced calcium signaling in *Trypanosoma cruzi*. Biochem Biophys Res Commun 281:300-4.

Catisti R, Uyemura SA, Docampo R, Vercesi AE. 2000. Calcium mobilization by arachidonic acid in trypanosomatids. Mol Biochem Parasitol 105:261-71.

Docampo R, Moreno SNJ. 1996. The role of Ca^{2+} in the process of cell invasion by intracellular parasites. Parasitol Today 12:61-5.

Docampo R, Moreno SNJ. 2001. The acidocalcisome. Mol Biochem Parasitol 33:151-9.

Flawiá MM, Tellez-Iñón MT, Torres HN. 1997. Signal transduction mechanisms in *Trypanosoma cruzi*. Parasitol Today 13:30-3.

Furuya T, Zhong L, Meyer-Fernandes JR, Lu H-G, Moreno SNJ, Docampo R. 1998. Ecto-protein tyrosine phosphatase activity in *Trypanosoma cruzi* infective stages. Mol Biochem Parasitol 92:339-48.

Furuya T, Kashuba C, Docampo R, Moreno SNJ. 2000. A novel phosphatidylinositol-phospholipase C of *Trypanosoma cruzi* that is lipid modified and activated during trypomastigote to amastigote differentiation. J Biol Chem 275:6428-38.

Furuya T, Okura M, Ruiz FA, Scott DA, Docampo R. 2001. TcSCA complements yeast mutants defective in Ca^{2+} pumps and encodes a Ca^{2+}-ATPase that localizes to the endoplasmic reticulum of *Trypanosoma cruzi*. J Biol Chem 276:32437-45.

Gomez ML, Ochatt CM, Kazanietz MG, Torres HN, Tellez-Iñón MT. 1999. Biochemical and immunological studies of protein kinase C from *Trypanosoma cruzi*. Int J Parasitol 29:981-9.

Grellier P, Blum J, Santana J, Bylen E, Mouray E, Sinou V, Teixeira AR, Schrevel J. 1999. Involvement of calyculin A-sensitive phosphatase(s) in the differentiation of *Trypanosoma cruzi* trypomastigotes to amastigotes. Mol Biochem Parasitol 98:239-52.

Kelley GG, Reks SE, Ondrako JM, Smrcka AV. 2001. Phospholipase Cε: a novel Ras effector. EMBO J 20:743-54.

Orr GA, Werner C, Xu J, Bennet M, Weiss LM, Takvorkan P, Tanowitz HB, Wittner M. 2000. Identification of novel serine/threonine protein phosphatases in *Trypanosoma cruzi*: a potential role in control of cytokinesis and morphology. Infect Immun 68:1350-8.

Parsons M, Ruben L. 2000. Pathways involved in environmental sensing in trypanosomatids. Parasitol Today 16:56-62.

Pascuccelli V, Labriola C, Tellez-Iñón MT, Parodi AJ. 1999. Molecular and biochemical characterization of a protein kinase B from *Trypanosoma cruzi*. Mol Biochem Parasitol 102:21-33.

Piacenza L, Peluffo G, Radi R. 2001. L-Arginine-dependent suppression of apoptosis in *Trypanosoma cruzi*: contribution of the nitric oxide and polyamine pathways. Proc Natl Acad Sci USA 98:7301-6.

Ruiz FA, Rodrigues CO, Docampo R. 2001. Rapid changes in polyphosphate content within acidocalcisomes in response to cell growth, differentiation, and environmental stress in *Trypanosoma cruzi*. J Biol Chem 276: 26114-21.

Taylor MC, Muhia DK, Baker DA, Mondragon A, Schaap P, Kelly JM. 1999. *Trypanosoma cruzi* adenylyl cyclase is encoded by a complex multigene family. Mol Biochem Parasitol 104:205-17.

Wilkowsky SE, Barbieri MA, Stahl P, Isola EL. 2001. *Trypanosoma cruzi*: phosphatidylinositol 3-kinase and protein kinase B activation is associated with parasite invasion. Exp Cell Res 264:211-8.

Docampo and Moreno

Zhong, L., Lu, H.-G., Moreno, S.N.J., Docampo, R. 1998.Tyrosine phosphate hydrolysis of host proteins by *Trypanosoma cruzi* is linked to cell invasion. FEMS Microbiol Lett 161:15-20.

GENETIC DIVERSITY OF *TRYPANOSOMA CRUZI*

O. Fernandes[*] and B. Zingales[**]
*Department of Tropical Medicine, Instituto Oswaldo Cruz, Fiocruz, Rio de Janeiro, Brazil.
**Instituto de Química, Universidade de São Paulo; São Paulo, Brazil

ABSTRACT

Chagas disease has a variable clinical presentation, and chronic manifestations have a disparate geographical distribution. As we describe here, *Trypanosoma cruzi* strains show great biological and genetic heterogeneity. Isoenzymes (MLEE), random amplification of polymorphic DNA (RAPD), mini-exon and ribosomal genes, all distinguish two major phylogenetic lineages, officially named *T. cruzi* I and *T. cruzi* II. Ribosomal spacer polymorphisms further discriminate subgroups; genetic fingerprinting and kinetoplast DNA (kDNA) polymorphisms give still more sensitive resolution. Remarkably, kDNA signatures of *T. cruzi* from heart and oesophagus of the same patient may differ. A major challenge remains the identification of *T. cruzi* molecular markers that could be used to predict clinical prognosis.

INTRODUCTION

On April 15, 1909 a day after having detected flagellates in the bloodstream of a two-year old girl, Carlos Chagas wrote a note announcing the discovery of a new human trypanosomiasis. The scientific skills of Carlos Chagas enabled him in a few years to describe the parasite named *Trypanosoma cruzi*, its development in triatomine vectors, the existence of sylvatic reservoirs and the clinical features of the disease in humans. Today, after extensive serological and clinical investigations carried out in many countries of Latin America it is recognised that Chagas disease has different clinical presentations: approximately 70% of the individuals are asymptomatic; chronic cardiopathy develops in 27% of infected people; digestive lesions in 6% and neurological disorders in 3%. Geographic differences in the prevalence of these presentations and in the response to specific treatment have been reported. Investigators of Chagas disease have been debating for a long time whether these differences are due to the parasite genotypes, the host genetic background, environmental factors or to the interplay of these elements.

It is widely known that *T. cruzi* is composed of a pool of organisms - named strains, stocks or isolates - that circulate among mammalian hosts and insect vectors. Early studies revealed that parasite strains show great heterogeneity in biological parameters such as morphology, antigenic make-up, virulence, pathogenicity and sensitivity to drugs.

An interesting characteristic of *T. cruzi* is that its genome size varies among the strains and clones. Early estimates of the DNA content range from 125 to 280 fg per cell of various strains, including the kinetoplast DNA (kDNA), which contributes 16-30% of the total DNA (Dvorak et al., 1982; McDaniel and Dvorak, 1993). The kinetoplast is a DNA-containing structure localised in the single mitochondrion of the parasite.

The significant differences in the DNA content suggested a possible correlation between the antigenic make-up of the strains and their biological characteristics. In this direction, a major challenge to the scientific community has been the identification of genetic markers of *T. cruzi* that could characterise particular groups of strains. At this point, it is important to mention that *T. cruzi* is primarily at least a diploid organism, with predominantly asexual reproduction and, therefore, its strains have usually been considered to represent independent clonal lineages (Tibayrenc et al., 1986).

DEFINING MOLECULAR MARKERS FOR *T. CRUZI* IDENTIFICATION AND TYPING

Each genome contains genetic markers with distinct rates of molecular evolution. The choice of genetic marker(s) for a given organism will depend on the desired objectives, including practical applications:

(a) for diagnostic purposes, the genetic target should be species-specific, preferably abundant and conserved among the different populations of the organism;

(b) for typing purposes, the molecular target should be able to discriminate populations of the same species. In this case, intergenic regions or DNA spacers have been preferentially used, since these sequences present a higher mutation rate in comparison with protein or RNA coding genes;

(c) for clustering strains into defined groups, it is of interest to employ genetic markers of intermediate discriminating power. The groups can then be characterised in terms of biological, medical or epidemiological parameters;

(d) an additional application of molecular markers is the reconstruction of the evolutionary history of the organism and of its strains.

In this chapter we will discuss some molecular markers described for *T. cruzi* and their practical applications.

ENZYME ELECTROPHORETIC PROFILES

The first studies on population genetics of *T. cruzi* were undertaken by the group of Michael Miles in the late seventies and involved the analysis of the electrophoretic profiles of isoenzymes (Miles et al., 1977). Isoenzymes have been considered the "gold standard" in population genetics, and have been used in the exploration of genetic diversity in microorganisms. Although isoenzyme profiles represent a phenotypic variation, their use in population genetics relies on the assumption that they reflect directly the variability of the genes coding these proteins. In fact, variations in the electrophoretic migration of a polypeptide can be obtained by a single nucleotide change at one locus. Miles and co-workers (1977), based on the electrophoretic profile of six enzymes of 17 *T. cruzi* isolates from São Felipe (Bahia state, Brazil), concluded that the isolates could be clustered into two different groups or zymodemes. The first zymodeme (Z1) consisted of isolates from opossums and triatomines of the sylvatic cycle, whereas the second (Z2) comprised isolates from humans and domestic animals. The authors expanded the group-classification after analysis of six human isolates from the Amazon region and about 250 isolates from several localities in Brazil. The groups Z1 and Z2 were confirmed and an additional group - Z3 - also related to the sylvatic transmission cycle, was characterised (Miles et al., 1978; Miles et al., 1980). These studies seemed very encouraging, as distinct zymodemes could be associated with different epidemiological transmission patterns, namely the domestic and sylvatic cycles. Ready and Miles (1980), using numerical taxonomy, suggested that Z1 is more related to Z3 than to Z2, reflecting the initial biological correlation of the transmission cycles with the zymodemes. Following this triple zymodeme classification, Flint et al. (1984) were able to define monoclonal antibodies, which reacted specifically with Z1, Z2 and Z3.

These data suggested that the zymodemes and their immunological classification would be a standard taxonomic parameter for *T. cruzi*. However, a few years later, Tibayrenc et al. (1986) analysed the electrophoretic patterns of 15 enzymes (instead of the six enzymes used hitherto) in 121 isolates from North, Central and South America and in this study, the authors established 43 different zymodemes. Despite the great intra-group heterogeneity, they consolidated the isolates into three distinct clusters: zymodemes 1 to 25, zymodemes 26 to 34 and zymodemes 35 to 43. Apparently, these clusters had no correlation with any epidemiological parameter. Later, the same authors associated the first cluster (zymodeme 1 to 25) with Z1 defined by Miles and co-workers (1977) and reaffirmed that the other zymodemes (26 to 43) showed a high level of genetic polymorphism (Tibayrenc et al., 1993). Saravia et al. *(*1987) reported the isoenzyme patterns of 54 Colombian *T. cruzi* isolates and found 15 different zymodemes. After a zymotaxonomic analysis, these groups were clustered into two specific zymodemes: Z1 and Z3. In this case Z1 isolates occurred in both the domestic and sylvatic cycles. Taken as a whole, the analysis of isoenzyme patterns disclosed the genetic and phenotypic heterogeneity of the parasite populations. On the other hand, it suggested in some localities a tendency of strain clustering in respect to the domestic and sylvatic transmission cycles.

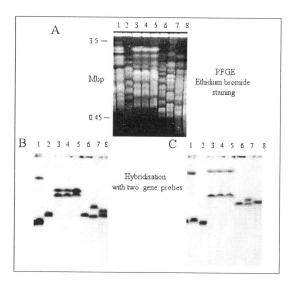

Figure 1. Molecular karyotype of *Trypanosoma cruzi* isolates. (A) chromosomal bands were separated by pulsed field gradient gel electrophoresis and DNA was stained with ethidium bromide. Two molecular size standards are indicated on the left. (B) DNA of the chromosomal bands was transferred to nylon membranes and hybridised with ^{32}P-labeled probes representing two independent genes. Samples: lanes 1 to 5, *T. cruzi* II strains (Z2), lanes 6 to 8, *T. cruzi* I strains (Z1).

KARYOTYPE ANALYSIS

Pulsed field gradient gel electrophoresis (PFGE) is a powerful technique in the study of the molecular karyotypes of protozoa, since in these organisms chromosomes do not condense during mitosis. PFGE reveals differences in chromosome sizes and chromosome number in *T. cruzi* strains (1A). The karyotype variability most probably reflects differences in the DNA content of the isolates, as described above.

Hybridisation of chromosomal bands with specific probes confirms the karyotype polymorphism, since the same probe may hybridise with chromosomes of different sizes in the various *T. cruzi* strains (Figure 1B). It has been proposed that a certain correlation exists between the strain karyotype and the zymodeme groups, and that Z2 isolates present chromosomes of larger molecular size when compared with Z1 (Henriksson et al., 1996; Zingales et al., unpublished results). This feature may reflect a speciation event and supports the distant relationship between zymodemes 1 and 2 (Ready and Miles, 1980). Karyotype data have been used for measuring genomic distances among populations of *Leishmania* species and *T. cruzi* groups (Dujardin et al., 2000).

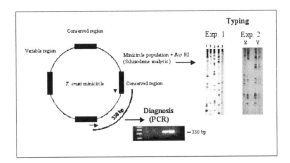

Figure 2. Schematic representation of *T. cruzi* minicircle molecule. Restriction fragment length polymorphism analysis of the kDNA network that comprises the whole minicircle population of a kinetoplast is shown on the right for typing purposes. In Experiment 1, kDNA extracted from five distinct strains isolated by xenodiagnosis from humans of the same endemic area was submitted to digestion with *Eco*RI. It can be observed that all patterns are different, depicting the diversity of the parasite in the same ecotope. Experiment 2 is a representation of the genetic diversity that can be found in the same Chagas disease patient. X and Y are strains from two different patients. These strains were isolated by xenodiagnosis. XA and YA were isolated using *Triatoma infestans,* XB and YB were isolated using *Panstrongylus megistus*. This reflects the potential role of the insect vector in selecting distinct populations of the parasite that circulate in the same human host. For diagnostic purposes, two primers (arrows) annealing in the conserved region of the minicircles (black rectangle) were employed to PCR-amplify the variable region. A 330bp product is obtained, which can be seen by agarose gel electrophoresis. This approach has been used for the molecular diagnosis of Chagas disease.

RESTRICTION FRAGMENT LENGTH POLYMORPHISM OF kDNA

The first breakthrough in the resolution of the high genetic variability in *T. cruzi* was the discovery of kinetoplast DNA (kDNA) restriction fragment length polymorphisms (RFLP). The RFLP approach involves the digestion of double-stranded DNA with restriction enzymes and analysis of the DNA fragments by gel electrophoresis. The number and sizes of the fragments resulting from the DNA digestion is called the restriction-fragment pattern.

The kDNA is an unusual structure localized in the single mitochondrion of Kinetoplastida and is composed of circular molecules named maxicircles and minicircles. The former contain genes that code for some of the mitochondrial proteins and tRNAs (Simpson, 1987). The minicircles code for small RNAs that are involved in the RNA editing process of maxicircle transcripts. In *T. cruzi,* the kinetoplast network is composed of approximately 50 maxicircles and 10,000 minicircles. The minicircle molecules of around 1,400bp contain four conserved regions that are separated by four variable regions (Degrave et. al., 1988; Figure 2). The variable regions have a higher mutation rate, which confers intra- and inter-minicircle sequence heterogeneity .

Morel et al. (1980) established the RFLP patterns of minicircle kDNA of several *T. cruzi* strains. The data confirmed the high intra-specific genetic variability, since characteristic patterns (called schizodemes) were obtained for different strains (Figure 2, Typing: Exp. 1). Very few population clusters with the same schizodeme could be found. The RFLP of kDNA was explored in different contexts and clearly showed that different parasite strains circulate in the same mammalian host. Moreover, that particular populations may be selected through distinct filters, such as haemoculture, xenodiagnosis, *in vitro* culture and infection of experimental animals (Morel and Simpson, 1980; Deane et al., 1984; Morel et al., 1986) (Figure 2, Typing: Exp. 2).

Schizodeme analysis requires considerable amounts of DNA, which means that trypanosomes from an infected host have to be cultured in axenic medium or inoculated into laboratory animals (Morel et al., 1984; Macina et al., 1987; Solari et al., 1991). This procedure may promote the selection of a particular parasite population, as discussed above. Through the development of recombinant DNA techniques, hyper-variable regions of minicircle kDNA of approximately 330bp could be cloned. These sequences were used as molecular probes in hybridization experiments and discriminated different groups of strains (Macina et al., 1987).

The polymerase chain reaction (PCR) improved the study of heterogeneity of *T. cruzi* populations, especially as this technique does not require the previous culture of the parasite. Using this approach, Sturm et al. (1989) amplified the variable region of the minicircles (Figure 2, PCR) and used the amplified products of 330bp (amplicons) as substrate for digestion with restriction endonucleases. This schizodeme-like experiment produced complex restriction patterns. Other groups, after amplifying the variable regions of the minicircles of several *T. cruzi* isolates, performed cross hybridisations with the amplicons, showing that this segment of the minicircle molecule has a high group-specificity (Avila et al., 1990; Breniere et al., 1992; Britto et al., 1995). The schizodeme analyses revealed that the nucleotide sequence of the 330bp variable portion of the kDNA minicircle molecule evolves rapidly enough to produce differences between individual isolates.

One of the major contributions arising from the structure of kDNA minicircles was the utilisation of these target molecules for the sensitive and specific parasitological diagnosis of *T. cruzi*. Indeed, a specific pair of oligonucleotides annealing to the conserved region of kDNA and flanking the variable region was designed (Figure 2). Under standardised PCR conditions, the amplification of a 330bp product is observed in all *T. cruzi* strains (Figure 2, Diagnosis). This allows the detection of one thousandth of the DNA content of a single parasite cell (Avila et al., 1990). This diagnostic assay has been tested and validated in distinct epidemiological settings and geographic regions (Avila et al., 1993; Britto et al., 1995).

DNA FINGERPRINTING

Further characterisation of the genetic identity and heterogeneity of the strains, this time exploring nuclear DNA variability, came from studies of DNA fingerprinting with multilocus core probes directed to families of minisatellites (Macedo et al., 1992). DNA fingerprinting resulted in approximately the same high level of strain discrimination as schizodeme analysis, whilst having the advantages of higher stability and greater operational simplicity.

RANDOMLY AMPLIFIED POLYMORPHIC DNA

Genetic polymorphism in several organisms has also been studied by the randomly amplified polymorphic DNA (RAPD) technique. In this approach, anonymous loci are PCR-amplified using arbitrary primers and the resultant data may be used to construct genetic trees. The study of *T. cruzi* isolates by RAPD analysis has shown high levels of similarity between strains belonging to the same zymodeme (Tibayrenc et al, 1993; Steindel et al., 1993). However, the genetic polymorphism disclosed by RAPD and isoenzyme profiles is not capable of producing the same resolution as schizodeme analysis and DNA fingerprinting.

MICROSATELLITES

The description of *T. cruzi* microsatellites consisting of cytosine-adenine repeats has provided a new tool for the analysis of the parasite population structure (Oliveira et al., 1998). These DNA markers are extremely polymorphic and dispersed throughout the parasite nuclear genome. Microsatellites have been employed to determine whether a *T. cruzi* isolate is a monoclonal or multiclonal population.

MINI-EXON AND RIBOSOMAL RNA GENES

Analyses of sequences that have a lower evolutionary rate, such as the ribosomal RNA genes (which are classical markers of evolution), and the mini-exon genes (which are of taxonomic value in trypanosomatids), indicated a clear dimorphism in *T. cruzi* strains. This is in contrast to the genetic hypervariability suggested by former typing approaches.

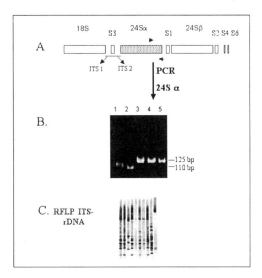

Figure 3. Ribosomal cistron and typing approaches. (A) Schematic representation of the ribosomal cistron of *Trypanosoma cruzi*. ITS, internal transcribed spacer, arrows, position of primers for the PCR amplification of a region of the 24Sα rRNA gene. (B) Typing of *T. cruzi* isolates by PCR amplification of 24Sα rRNA gene. Samples: lane 1, group 1/2, lane 2, group 2, lanes 3 to 5, group 1. (C) RFLP of ITS-rDNA. The region comprising ITS1-S3-ITS2 of different isolates was amplified by PCR and digested with restriction enzymes. The products were analysed by acrylamide gel electrophoresis and stained with ethidium bromide.

Ribosomal cistron

In *Trypanosoma* species, the ribosomal cistron exhibits a usual organisation (Figure 3A). Souto and Zingales (1993) and Souto et al. (1996) demonstrated that a sequence of approximately 100bp located at the 3' end of the 24Sα ribosomal RNA gene is dimorphic and can be used as a target for strain typing. PCR amplification with primers D71 and D72 (see Figure 3A) gives products of different sizes, which allow classification of strains into three groups, as follows: group 1, 125bp product; group 2, 110bp product; group 1/2, both products (Figure 3B). The sequence of the 18S rDNA gene has been also used for typing analyses. Riboprinting of the 18S rDNA, a technique which involves digestion of PCR amplified products with a battery of restriction enzymes, defined two or three distinct patterns, - named ribodemes 1, 2 and 3 - that correlated with the host from which the stocks were isolated (Clark and Pung, 1994).

Mini-exon gene

The mini-exon gene is present in trypanosomatid genomes as 100-200 tandem repeats and encodes a 39-nucleotide exon that is added post-transcriptionally to the 5' end of all mRNAs. The mini-exon gene repeats are separated by non-transcribed spacers that show genus- or species-specific heterogeneity. After a detailed study of the sequences of the mini-exon intergenic regions of different *T. cruzi* isolates, it was possible to detect some level of heterogeneity (Souto et al., 1996). Then, specific primers were designed, targeted to a hypervariable region for molecular genotyping approaches (Figure 4A and B). Approximately 90 stocks from different hosts and regions of South America were simultaneously typed by PCR amplification of the 24Sα rDNA and the mini-exon gene non-transcribed spacer. A perfect correlation between the two genotyping approaches was observed indicating the presence of two major clusters in the taxon *T. cruzi* (Souto et al., 1996). Additional analysis of 50-60 anonymous loci by RAPD allowed the definition of two major lineages presenting a high phylogenetic divergence. The two lineages broadly correspond to zymodemes Z2 and Z1, determined by Miles, encompassing strains of 24Sα rDNA groups 1 and 1/2 (Z2) and strains of rDNA group 2 (Z1) (Souto et al., 1996).

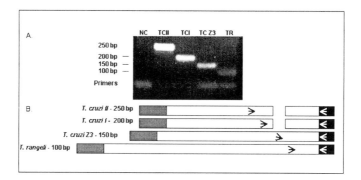

Figure 4.   Mini-exon multiplex PCR assay. (A). Agarose gel electrophoresis of amplified product of reference strains of *T. cruzi* I (TCI), T. *cruzi* II (TCII), *T. cruzi* Z3 (TCZ3) and *T. rangeli* (TR). NC. Negative control where no DNA was added to the reaction. (B). Amplification scheme showing the annealing site of the primers (arrows) and the size of the amplification products. Boxes represent the approximate sizes of the mini-exon genes of *T. cruzi* I, II, Z3 and *T. rangeli.* The black boxes correspond to the conserved 39 nucleotide exon. The introns are represented by the grey boxes. The open boxes correspond to the non-transcribed spacers of the mini-exon gene.

Epidemiological surveys in several Brazilian states as well as in Bolivia indicate that the lineage corresponding with Z2 predominates in the domestic cycle, being isolated mostly from chagasic patients, whereas the lineage corresponding with Z1 is preferentially encountered in the sylvatic cycle (Fernandes et al., 1998a; Zingales et al., 1998). Nevertheless, Z1 occurs in both domestic and silvatic cycles in Venezuela and other countries North of the Amazon. The PCR typing approach has a clear advantage over determination of isoenzyme patterns since it can be performed on a very low number of parasites.

Comparative sequence analysis of the amplified products of the 24Sα rDNA indicated that the divergence between the two *T. cruzi* lineages occurred before the divergence of the extant strains (Briones et al., 1999). Phylogenetic reconstruction using the 18S rRNA gene suggests that the divergence time of the two lineages is between 88 and 37 million years ago (Briones et al., 1999). Different genome targets may display distinct estimates for time evolution in *T. cruzi* and even more significant population clusters may be found (Machado and Ayala, 2001).

The third isoenzymic group of *T. cruzi*, named zymodeme 3 (Z3) (Miles et al., 1978), was described in the Amazon region, associated with the sylvatic transmission cycle, involving armadillos, terrestrial marsupials, triatomine species and rarely circulating among humans (Barrett et al., 1980; Miles et al., 1981; Povoa et al., 1984). The mini-exon gene of zymodeme 3 isolates presents a peculiar insertion in the non-transcribed spacer flanked by direct repeats. This observation raised the hypothesis of homogolous recombination processes determining the diversity of this tandem repeated gene in *T. cruzi* (Fernandes et al., 1998b; Fernandes et al., 2001).

Although the mini-exon and the 24Sα rDNA markers are suitable for defining the two major phylogenetic groups of *T. cruzi*, these targets do not allow characterisation of intra-group polymorphism since both DNA sequences are slowly evolving. On the other hand, ribosomal spacers are suitable markers for intra-group discrimination.

An internal transcribed spacer (ITS1) separates the coding region of the 18S rDNA and the S3 rDNA. Another ITS (ITS2) separates the S3 rDNA from the 24Sα rDNA (Figure 3A). The analysis of restriction fragment length polymorphism (RFLP) of the ITS-rDNA (ITS1 + S3 rDNA + ITS2) (3C) has been used to establish the genetic relationship among *T. cruzi* isolates. The phenetic dendrogram resultant from RFLP profiles clustered several isolates into the two *T. cruzi* major phylogenetic lineages (Fernandes et al., 1999) and showed that *T. cruzi* Z3 stocks derived from the Amazon region are divided into two subgroups.

LOW-STRINGENCY SINGLE-SPECIFIC PRIMER-POLYMERASE CHAIN REACTION (LSSP-PCR)

To unravel the molecular epidemiology of Chagas disease at a fine level and to try to understand why different patients develop cardiac, digestive or indeterminate clinical forms, the development of typing techniques that allow the strain characterisation in infected tissues was undertaken. Thus, a sensitive DNA profiling technique called low-stringency single-specific primer-polymerase chain reaction (LSSP-PCR) was standardised, which is directed to the variable region of kDNA. The intra-specific polymorphism of kDNA sequence can be translated into individual and highly reproducible complex "kDNA signatures" (Vago et al., 1996; Vago et al., 2000). The LSSP-PCR technique has been used in profiling the *T. cruzi* parasites present in the hearts of patients with chagasic cardiomyopathy and in the oesophagus of patients with digestive disorders. The data clearly show a different kDNA signature in each case, which reflects the great genetic variability in *T. cruzi*. An interesting finding was the observation that the kDNA signatures obtained from the heart and oesophagus of the same patient differed significantly in the two organs. This suggests that the same individual was infected by a multiclonal strain and that one or more clone(s) would have different tissue/organ tropism (Vago et al., 2000).

GENETIC EXCHANGE

Mitotic division (clonal propagation or asexual multiplication) is the predominant way of reproduction in *T. cruzi* as demonstrated by the Hardy-

Weinberg equilibrium and linkage disequilibrium. However, genetic hybrids characterized by multilocus enzyme electrophoresis and RAPD have been described circulating sympatrically with putative parents. Experimental hybrids, displaying shared parental phenotypic and genotypic characters, have been generated in the laboratory, from genetically transformed biological clones. Studies based on dual drug-resistant *T. cruzi* biological clones showed that the hybrids presented both episomal drug-resistant markers from the single-drug resistant parents that were co-cultured (Stothard et al., 1999).

In conclusion, genetic exchange should be considered in *T. cruzi* as a phenomenon generating hybrid phenotypes (multilocus enzyme electrophoresis) and genotypes (RAPD, nuclear and mitochondrial sequence analysis) as described above. Recently, Machado and Ayala (2001), based on nuclear (DHFR-TS and trypanothione reductase) and mitochondrial sequences (maxicircle-encoded genes cytochrome oxidase subunit II and NADH dehydrogenase subunit I), described evidence of hybridization between strains from two divergent groups of *T. cruzi*, suggesting that genetic exchange should be considered as having a major role in generating the genetic diversity in the parasite taxon.

CONCLUDING REMARKS

Several molecular typing approaches clearly indicate the division of *T. cruzi* into two major phylogenetic lineages. These lineages were recently officially named *T. cruzi* I and *T. cruzi* II (Satellite Meeting, 1999). The general epidemiological characteristics of the two groups have been established. So far genetic markers do not, however, allow reliable discrimination of strains inducing different clinical manifestations of Chagas disease. Typing techniques, such as LSSP-PCR, give very complex patterns for strains obtained from heart and oesophagus of chagasic individuals. A major challenge is the identification of molecular markers which present intermediate polymorphism of *T. cruzi* strains, and that could be used as a prognostic tool for the evolution of a particular clinical presentation. In addition, it is of importance to determine the contribution of host genetic differences to the variable development of Chagas disease.

Parasite genome projects have been undertaken aiming at improving disease control. The *T. cruzi* genome project was launched in 1994, employing as the reference organism the CL Brener clone, which belongs to *T. cruzi* II. Bearing in mind the extensive genetic diversity of *T. cruzi* strains, comparative genomics between different *T. cruzi* strains will be highly relevant.

References

Avila HA, Gonçalves AM, Nehme NS, Morel CM, Simpson L. 1990. Schizodeme analysis of *Trypanosoma cruzi* stocks from South and Central America by analysis of PCR-amplified minicircle variable region sequences. Mol Biochemical Parasitol 12:175-188.

Avila HA, Pereira JB, Thiemann O, Paiva E, Degrave W, Morel CM, Simpson L. 1993. Detection of *Trypanosoma cruzi* in blood specimens of chronic chagasic patients by polymerase chain reaction amplification of kinetoplast DNA: comparison with serology and xenodiagnosis. J Clin Microbiol 31:2421-2426.

Barrett TV, Hoff RH, Mott K, Miles MA, Godfrey DG, Teixeira R, Almeida de Souza JA, Sherlock IA. 1980. Epidemiological aspects of three *Trypanosoma cruzi* zymodemes in Bahia State, Brazil. Trans R Soc Trop Med Hyg 74:84-90.

Brenière SF, Bosseno MF, Revollo S, Rivera MT, Carlier Y, Tibayrenc M. 1992. Direct identification of *Trypanosoma cruzi* natural clones in vectors and mammalian hosts by polymerase chain reaction amplification. Am J Trop Med Hyg 46:335-341.

Briones MRS, Souto RP, Stolf BS, Zingales B. 1999. The evolution of two *Trypanosoma cruzi* subgroups inferred from rRNA genes can be correlated with the interchange of American mammalian faunas in the Cenozoic and has implications to pathogenicity and host specificity. Mol Biochem Parasitol 104:219-232.

Britto C, Cardoso MA, Monteiro Vanni CM, Hasslocher-Moreno A, Xavier SS, Oelemann W, Santoro A, Pirmez C, Morel CM, Wincker P. 1995. Polymerase chain reaction detection of *Trypanosoma cruzi* in human blood samples as a tool for diagnosis and treatment evaluation. Parasitology 110:241-247.

Clark CG, Pung OJ. 1994. Host specificity of ribosomal DNA variation in sylvatic *Trypanosoma cruzi* from North America. Mol Biochem Parasitol 66:174-179.

Deane MP, Sousa MA, Pereira NM, Gonçalves AM, Momen H, Morel CM. 1984. *Trypanosoma cruzi*: inoculation schedules and re-isolation methods select individual strains from doubly infected mice, as demonstrated by schizodeme and zymodeme analyses. J Protozool 31:276-280.

Degrave W, Fragoso S, Britto C, Van Heuverswyn H, Kidane G, Cardoso M, Mueller R, Simpson L, Morel C. 1988. Peculiar sequence organization of kinetoplast minicircles from *Trypanosoma cruzi*. Mol Biochem Parasitol 27:63-70.

Dujardin JC, Henriksson J, Victoir K, Brisse S, Gamboa D, Arevalo J and Ray D. 2000. Genomic rearrangements in trypanosomatids: an alternative to the "one gene" evolutionary hypothesis? Mem Inst Oswaldo Cruz 95:527-534.

Dvorak JA, Hall TE, Crane MSJ, Engel JC, McDaniel JP, Uriegas R. 1982. *Trypanosoma cruzi*: Flow cytometric analysis. I. Analysis of total DNA/organism by means of mithramycin - induced fluorescence. J Protozool 29:430-437.

Fernandes O, Souto RP, Castro JA, Pereira JB, Fernandes NC, Junqueira ACV, Naiff RD, Barret TV, Degrave W, Zingales B, Campbell DA, Coura JR. 1998a. Brazilian isolates of *Trypanosoma cruzi* from human and triatomines classified into two lineages using mini-exon and ribosomal RNA sequences. Am J Trop Med Hyg 58:807-811.

Fernandes O, Sturm NR, Derré R, Campbell DA. 1998b. The mini-exon gene: a molecular marker for zymodeme III of *Trypanosoma cruzi*. Mol Biochem Parasitol 95:129-133.

Fernandes O, Santos SS, Junqueira ACV, Jansen AM, Cupolillo E, Campbell DA, Zingales B, Coura JR. 1999. Populational heterogeneity of Brazilian *Trypanosoma cruzi* isolates revealed by the mini-exon and ribosomal spacers. Mem Inst Oswaldo Cruz 94:195-198.

Fernandes O, Santos SS, Cupolillo E, Mendonça B, Derre R, Junqueira ACV, Santos LC, Sturm NR, Naiff RD, Barrett TV, Campbell DA, Coura JR. 2001. A mini-exon multiplex polymerase chain reaction to distinguish the major groups of *Trypanosoma cruzi* and *T. rangeli* in the Brazilian Amazon. Trans R Soc Trop Med Hyg 95:1-3.

Flint JE, Schechter M, Chapman MD, Miles MA. 1984. Zymodeme and species specificities of monoclonal antibodies raised against *Trypanosoma cruzi*. Trans R Soc Trop Med Hyg 78:193-202.

Henriksson J, Aslund L, Petterson U. 1996. Karyotype Variability in *Trypanosoma cruzi*. Parasitol Today 12:108-114.

Macedo AM, Martins MS, Chiari E, Pena SDJ. 1992. DNA fingerprinting of *Trypanosoma cruzi*: a new tool for characterizatrion of strains and clones. Mol Biochem Parasitol 55:147-154.

Machado CA, Ayala FJ. 2001. Nucleotide sequences provide evidence of genetic exchange among distantly related lineages of *Trypanosoma cruzi*. Proc Natl Acad Sci USA 98:7396-7401.

Macina RA, Arauzo S, Reyes MB, Sanchez DO, Basombrio MA, Montamat EE, Solari A, Frasch ACC. 1987. *Trypanosoma cruzi* isolates from Argentina and Chile grouped with the aid of DNA probes. Mol Biochem Parasitol 25:45-53.

McDaniel JP Dvorak JA. 1993. Identification, isolation, and characterization of naturally-occurring *Trypanosoma cruzi* variants. Mol Biochem Parasitol 57:213-222.

Miles MA, Toyé PE, Oswald SC, Godfrey DG. 1977. The identification by isoenzyme patterns of two distinct strain-groups of *Trypanosoma cruzi*, circulating independently in a rural area of Brazil. Trans R Soc Trop Med Hyg 71:217-225.

Miles MA, Souza A, Povoa M, Shaw JJ, Lainson R, Toyé PE. 1978. Isozymic heterogeneity of *Trypanosoma cruzi* in the first authochthonous patients with Chagas disease in Amazonian Brazil. Nature 272:819-821.

Miles MA, Lanham SM, Souza AA, Póvoa M. 1980. Further enzymic character of *Trypanosoma cruzi* and their evaluation for strain identification. Trans R Soc Trop Med Hyg 74:221-237.

Miles MA, Povoa MM, Souza AA, Lainson R, Shaw JJ, Ketteridge DS. 1981. Chagas disease in the Amazon Basin: II. The distribuition of *Trypanosoma cruzi* zymodemes 1 and 3 in Para State, North Brazil. Trans R Soc Trop Med Hyg, 75:667-674.

Morel CM, Simpson, L. 1980. Characterization of pathogenic Trypanosomatidae by restriction endonuclease fingerprint of kinetoplast DNA minicircle. Am J Trop Med Hyg 29:1070-1074.

Morel CM, Chiari E, Camargo EP, Mattei DM, Romanha AJ, Simpson, L. 1980. Strains and clones of *Trypanosoma cruzi* can be characterized by pattern of restriction endonuclease products of Kinetoplast DNA mincircles. Proc Natl Acad Sci USA 77:6810-6814.

Morel CM, Gonçalves AM, Simpson L, Simpson A. 1984. Recent advances in the development of DNA hybridization probes for the detection and characterization of *Trypanosoma cruzi.* Mem Inst Oswaldo Cruz 79:51-53.

Morel CM, Deane MP, Gonçalves AM. 1986. The complexity of *Trypanosoma cruzi* populations revealed by schizodeme analysis. Parasitol Today, 4:97-101.

Oliveira RP, Broude NE, Macedo AM, Cantor CR, Smith CL, Pena SDJ. 1998. Probing the genetic population structure of *Trypanosoma cruzi* with polymorphic microsatellites. Proc Natl Acad Sci USA 95:3776-3780.

Povoa MM, De Souza AA, Naiff RD, Arias JR, Naiff MF, Biancaardi CB, Miles MA. 1984. Chagas disease in the Amazon Basin IV. Host records of *Trypanosoma cruzi* zymodemes in the states of Amazon and Rondônia, Brazil. Ann Trop Med Parasitol 78:479-487.

Ready PD, Miles MA. 1980. Delimitation of *Trypanosoma cruzi* zymodemes by numerical taxonomy. Trans R Soc Trop Med Hyg 74:238-241.

Saravia NG, Holguín AF, Cibulskis RE, D'Alessandro A. 1987. Divergent isoenzyme profiles of sylvatic and domiciliary *Trypanosoma cruzi* in the eastern plains, piedmont, and highlands of Colombia. Am J Trop Med Hyg 36:59-69.

Satellite meeting 1999. Recommendations from an International Symposium. Mem Inst Oswaldo Cruz 94:429-432.

Simpson L. 1987. The mitochondrial genome of kinetoplastid protozoa: genomic organization, transcription, replication and evolution. Annu Rev Microbiol 41:363-382.

Solari A, Venegas J, Gonzalez E, Vasquez C. 1991. Detection and classification of *Trypanosoma cruzi* by DNA hybridization with non-radioactive probes. J Protozool 38:559-565.

Souto R, Zingales B. 1993. Sensitive detection and strain classification of *Trypanosoma cruzi* by amplification of a ribosomal RNA sequence. Mol Biochem Parasitol 62:45-52.

Souto RP, Fernandes O, Macedo AM, Campbell D, Zingales B. 1996. DNA markers define two major phylogentic lineages of *Trypanosoma cruzi* Mol Biochem Parasitol 83:141-152.

Steindel M, Dias Neto E, Menezes CLP, Romanha AJ, Simpson AJG. 1993. Random amplified polymorphic DNA analysis of *Trypanosoma cruzi* strains. Mol Biochem Parasitol 60:71-80.

Stothard JR, Frame IA, Miles MA. 1999. Genetic diversity and genetic exchange in *Trypanosoma cruzi*: dual drug-resistant "progeny" from episomal transformants. Mem Inst Oswaldo Cruz 94 Suppl. 1:189-193.

Sturm NR, Degrave W, Morel CM, Simpson L. 1989. Sensitive detection and schizodeme classification of *Trypanosoma cruzi* cells by amplification of kinetoplast minicircle DNA sequences: use in diagnosis of Chagas disease. Mol Biochem Parasitol 33:205-214.

Tibayrenc M, Ward P, Moya A, Ayala FJ. 1986. Natural populations of *Trypanosoma cruzi*, the agent of Chagas disease, have a complex multiclonal structure. Proc Natl Acad Sci USA, 83:115-119.

Tibayrenc M, Neubauer K, Barnabé C, Guerrini F, Skarecky D, Ayala FJ. 1993. Genetic characterization of six parasitic protozoa: parity between random-primer DNA typing and multilocus enzyme electrophoresis. Proc Natl Acad Sci USA 90:1335-1339.

Vago AR, Andrade LO, Leite AA, Reis AD, Macedo AM, Adad SJ, Tostes S, Moreira MCV, Brasileiro G, Pena SDJ. 2000. Genetic characterization of *Trypanosoma cruzi* directly from tissues of patients with chronic Chagas disease. Am J Pathol 156:1805-1809.

Vago AR, Macedo AM, Oliveira RP, Andrade LO, Chiari E, Galvão LMC, Reis DA, Pereira MES, Simpson AJG, Tostes S, Pena SDJ. 1996. kDNA signatures of *Trypanosoma cruzi* strains obtained directly from infected tissues Am J Pathol 149:2153-2159.

Zingales B, Souto R, Mangia R, Lisboa C, Campbell D, Coura J, Jansen A, Fernandes O. 1998. Molecular epidemiology of American Trypanosomiasis in Brazil based on dimorphisms of rRNA and mini-exon gene sequences. Int J Parasitol 28:105-112.

FUNCTIONAL DISSECTION OF THE *TRYPANOSOMA CRUZI* GENOME: NEW APPROACHES IN A NEW ERA

M. C. Taylor and J. M. Kelly
Pathogen Molecular Biology and Biochemistry Unit, Department of Infectious and Tropical Diseases, London School of Hygiene and Tropical Medicine, Keppel Street, London WC1E 7HT UK

ABSTRACT

The study of trypanosomatids has led to the discovery of many novel genetic and biochemical phenomena. This has had a major impact on our understanding of disease pathogenesis and the mechanisms used by the parasites to avoid immune destruction. Central to these investigations has been the development of procedures for trypanosomatid genetic manipulation. In the current chapter we will outline the range of techniques that can now be applied to *Trypanosoma cruzi* and describe some of the important questions that have been addressed. We will also speculate on the likely direction of future research given the rapid progress of the *T. cruzi* genome project and the increasing availability of emerging high-throughput technologies.

THE STRANGE GENOME ORGANISATION OF *T. CRUZI*

T. cruzi is an extremely divergent species and displays considerable variation at the level of chromosome structure and organisation. The genome is diploid and nuclear DNA content can range from 50-100 million base pairs (Mbp), depending on the strain. Estimates for the gene copy suggest that there are 10-14,000 genes. Trypanosome chromosomes do not condense during mitosis and therefore karyotypic studies have relied on pulsed field gel electrophoresis and associated techniques. These indicate that the genome is organised into at least 20 chromosome pairs ranging from 0.4 to 4.5 Mbp, although chromosome homologues are often very different in size. Originally this was thought to reflect differences in the number of repeat sequence elements, particularly in the sub-telomeric domains. However, more recent analysis has suggested that large insertions and/or deletions in the internal regions of chromosomes also make a major contribution to this size variation. Analysis of genome structure has been further complicated by reports of genetic exchange in *T. cruzi* (Stothard et al., 1999; Machado and Ayala, 2001; Miles et al., in press). Although infrequent, this appears to have made a significant contribution to genetic diversity. Phylogenetically, the CL Brener clone that has been selected for genome sequencing has been described as *T. cruzi II* (See Chapter on Genetic Diversity, this volume). Recent evidence, however, suggests that the CL Brener is actually a hybrid of two distantly related lineages or Clades (B and C) within *T. cruzi II* (Machado and Ayala, 2001). This has two major implications for genome analysis. Firstly, allelic copies of genes can vary in sequence by at least 1.5%. Secondly, due to recombination each chromosome may represent a mosaic of the parental genotypes.

Approximately 35% of the *T. cruzi* genome is made up of repetitive sequences (Aguero et al., 2000). These include the SIRE elements, minisatellites and sequences of a retrotransposon origin. In addition, many house-keeping genes are organised into large tandemly repeated arrays. For example, the genes encoding cruzipain, the major cysteine proteinase, are organised into large clusters, with up to 100 copies in some strains. The genes that encode the surface glycoproteins form large heterogenous families and are widely dispersed in the genome. These include members of the trans-sialidase super-family, such as gp90, gp85 and gp82, and a family of about 500 genes that encode the variant mucin-like molecules that are the major glycoproteins on the cell surface.

One of the most striking features of trypanosomatids is the organisation of the mitochondrial genome. These organisms contain a single mitochondrion, which in *T. cruzi* can form a large complex structure that extends throughout much of the cell. The mitochondrial DNA is localised to a specific site within the organelle and forms a tight compact disc called the kinetoplast, which is situated adjacent to the flagellar pocket. The kinetoplast contains two types of circular DNA molecules that form a concatenated network. The maxicircle DNA molecules, analogous to the mitochondrial genome in other organisms, are between 20-30 kb long and are present in 25-50 copies per cell. Maxicircle DNA contains genes coding for ribosomal RNA and some components of the mitochondrial respiratory system. Most unusually, many of the primary transcripts of the protein coding genes do not contain a complete open reading frame and are required to undergo a process of RNA editing to produce functional mRNA. This maturation event involves the post-transcriptional insertion or deletion of uridine residues. The second type of DNA molecules present in the mitochondrion are the minicircles, which are present in 5-10,000 copies and have a size range of 0.5-2.5 kb. Minicircles do not contain protein coding genes, but instead encode a class of small 3'-oligouridylated transcripts known as guide RNAs which have a role in RNA editing. Minicircles have a conserved region, found in all trypanosomatids, which encompasses the postulated origin of replication and a region of high variability that reflects the repertoire of expressed guide RNAs. Recently, considerable progress has been made in dissecting the mechansims of RNA editing (Schnaufer et al., 2001; Simpson et al., 2000)

GENE EXPRESSION IN TRYPANOSOMATIDS: AN UNUSUAL PROCESS

In most eukaryotes, the binding of transcription factors to promoters that are located upstream of each gene facilitates recruitment of the transcription complex. For protein coding genes transcription is mediated by RNA polymerase II (pol II). The primary transcripts undergo a series of tightly regulated post-transcriptional processing events that include capping, removal of introns by *cis*-splicing and polyadenylation before they are exported to the cytoplasm. In trypanosomatids, the mechanisms of gene expression and post-transcriptional processing differ in almost every way from those in other eukaryotes. Pol II promoters for protein coding genes have yet to be identified in *T. cruzi*, or in other trypanosomatids. Instead linked genes are co-transcribed in long polycistronic units, often extending over several hundred thousand nucleotides (Figure 1). Preliminary mapping and sequencing of trypanosomatid chromosomes indicates that transcription can proceed from the centre of the chromosome towards the telomeres in some cases, and vice-versa in others. The nature of the transcription initiation sites and the extent to which they are regulated remains to be determined. Because the steady state levels of mRNA derived from linked genes can often differ by several hundred-fold, it is implicit that regulation must be predominantly a post-transcriptional event.

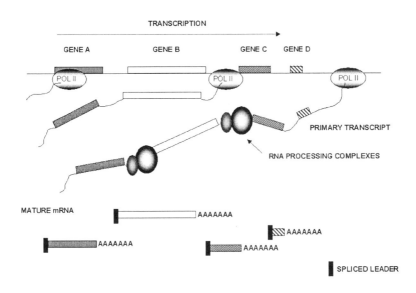

Figure 1. Polycistronic transcription in trypanosomes: In trypanosomes linked genes are generally arranged in the same orientation and transcribed as a large primary transcript. In this example genes A-D are linked along the chromosome. The transcription complex transcribes all four into one primary transcript. The transcript is then processed by the *trans*-splicing of a 39 nucleotide spliced leader into the 5' end of each coding sequence and polyadenylation at the 3'end to yield the mature translatable mRNA. The polyadenylation of the upstream mRNA is coupled to the *trans*-splicing of the downstream mRNA via interaction of the RNA processing complexes with a polypyrimidine tract found in the intergenic sequence.

In trypanosomatids mature mRNAs are generated by *trans*-splicing, a reaction that involves the joining of exons from two separately transcribed molecules, the spliced leader (SL) RNA and pre-mRNAs (Figure 1). The SL RNAs are encoded by the so-called 'mini-exon' genes, which are present in approximately 100-200 tandemly repeated copies. These transcripts contribute the 39 nucleotide sequence, including a cap structure of 4 modified nucleotides, that forms the 5'-end of mature mRNA (Figure 1). Addition of the SL sequence to the pre-mRNA involves the formation of an intermediary Y branched structure, analogous to the lariat structure formed during *cis*-splicing in other eukaryotes. The process is catalysed by splicing machinery that bears strong similarities to that used for *cis*-splicing. *Trans*-splicing has also been shown to be functionally coupled to polyadenylation of the upstream transcript. Addition of the poly-A tail does not require a conventional consensus sequence such as the polyadenylation signal (AATAAA) of other eukaryotes. Rather, polypyrimidine rich sequences seem to be an important element for specifying the addition site of the SL RNA and the poly-A tail, although both can occur at several alternate places within a window. To date the only intron found in *T. cruzi* is localised in the poly-A polymerase gene.

Regulation at the level of mRNA stability is of major importance in *T. cruzi*, particularly for genes that are expressed differentially during the life-cycle. Control often appears to be mediated by sequences in the 3'-untranslated region (UTR) of the mRNA. These act to stabilise, or destabilise, the transcript in a stage-specific manner. Examples include the genes encoding amastin, gp85 and the mucin-like glycoproteins. In the latter case, AU-rich sequence elements in the 3'-UTR and their interaction with differentially expressed RNA binding proteins appear to be of crucial significance (D'Orso and Frasch, 2001). Translational and post-translational mechanisms have also been shown to have a major role in regulating gene expression in *T. cruzi*. For example, although the level of cruzipain mRNA is constant throughout the life-cycle, enzyme activity is 5 times higher in the insect stage than in the trypomastigote and amastigote stages (Tomas and Kelly, 1996). In addition, whereas cruzipain is restricted to vesicles of the endosomal/lysosomal system in epimastigotes, the enzyme is readily detectable on the cell surface of the intracellular forms.

THE DEVELOPMENT OF GENETIC TRANSFORMATION TECHNIQUES FOR *T. CRUZI*.

Stable transformation techniques applicable to parasitic trypanosomatids have been available for more than 10 years (Clayton, 1999; Kelly, 1997). These procedures have been central to the functional analysis of parasite genes. They have had a major role in the identification and validation of novel targets for chemotherapy, in the characterisation of genes crucial for parasite development and virulence, in the analysis of host-parasite interactions, and in unravelling the mechanisms of gene expression. The stable transformation of a parasitic protozoan was first demonstrated with *Leishmania*, followed soon after by *T. brucei* and *T. cruzi*. Electroporation was found to be the most efficient method of transfection. This procedure involves the use of short electric pulses to transiently permeabilise membranes enabling the cells to take up exogenous DNA. Integration of this DNA into the parasite genomes is exclusively by homologous recombination and gene replacement can readily be achieved. The integration constructs are usually designed so that the targeting DNA fragments are placed either side of a cassette containing a drug-selectable marker (often *neo^r* or *hyg^r*) flanked by 5'- and 3'- sequences that facilitate transcript processing (Figure 2A). Fragments of a few hundred nucleotides are sufficient to mediate targeting.

In the case of *T. cruzi* and *Leishmania*, the construction of episomal shuttle vectors that allow the expression of transfected genes in transformed cells has also been reported (Figure 2B) (Kelly, 1995). The uses of these vectors include the ability to overexpress endogenous genes, to express modified genes or antisense RNA, or to express genes at inappropriate stages of the life-cycle. The mechanisms of replication and transcription initiation in these constructs have not been established. The plasmids undergo random segregation in transfected cells and often form large concatamers. Parasites with a high plasmid copy number (often in the range 50-100) can be selected following growth at increased drug concentrations. By modifying plasmid expression vectors such as pTEX (Figure 2B) with the insertion of a ribosomal RNA promoter, it has also been possible to increase the level of expression of the selectable marker gene and reduce the time required for the selection of transformants (Martinez-Calvillo et al., 1997). In addition to plasmids, cosmid shuttle vectors have also been developed for work on *T. cruzi* and *Leishmania*. Cosmids are modified plasmids that carry the *cos* sequences that function in the packaging of bacteriophage λ.

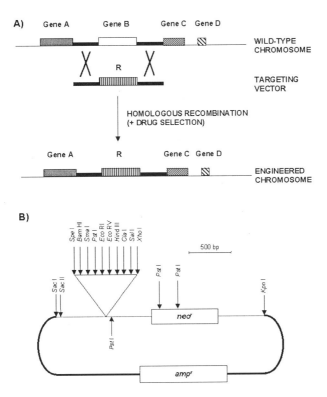

Figure 2.
A) Gene deletion in trypanosomes:
To target gene B the 5' and 3' flanking sequences (▬▬) are cloned on either side of a drug resistance marker (R). This fragment is electroporated into the trypanosomes where the flanks recombine with the endogenous locus resulting in the deletion of the Gene B coding sequence. Trypanosomes in which this has occurred are resistant to the antibiotic selection and so grow out when the wild-type cells die.
B) Expression vector pTEX
The pTEX expression vector can be used to express transfected genes in *T. cruzi* and *Leishmania* and is typical of the type vectors used for work on these organisms. The multiple cloning site (MCS) and the *neo^r* gene are flanked by the 5'-upstream and 3'-downstream regions of the *T. cruzi* glyceraldehyde-3 phosphate dehydrogenase genes (thin line). These sequences serve to ensure correct processing of the mRNA. The thick line, including the ampicillin resistance gene (*amp^r*), corresponds to sequences derived from the bacterial plasmid pBluescript. Genes cloned into the MCS can be expressed at high levels in transfected parasites.

   This allows these vectors to be used to clone DNA inserts in the range of 30–45 kb. Cosmids can readily be introduced into the parasites by electroporation, where they replicate as episomes in multiple copies. In *T. brucei* stable transformation with episomal vectors has proved more difficult to achieve. Integration is the favoured mechanism of transformation, the requirements for autonomous replication appear to be more stringent and expression of genes from episomes is dependent on the presence of a pol I, or pol I-like promoter.

EXTENDING THE RANGE AND SCOPE OF GENETIC TOOLS

The development of additional genetic tools and procedures has continued at a rapid pace. By modifying expression vectors it has been possible to use an epitope tagging approach for protein localisation studies. This has been achieved by engineering the episomal vector so that the expressed protein contains a short peptide tag identifiable with a specific antibody. In combination with fluorescence or immunoelectron microscopy this has proved to be rapid and powerful procedure for identifying the subcellular location of parasite proteins. Examples include the use of a 10 amino acid carboxyl-terminal tag derived from the human c-myc protein (Tibbetts et al., 1995) to demonstrate the mitochondrial sequestration of a 70 kDa heat shock protein and a member of the peroxiredoxin family of antioxidant proteins. As an alternative approach to investigate subcellular localisation, proteins, or defined regions of proteins, can be expressed attached to a marker such as the green fluorescent protein (GFP) of the jelly fish *Aequorea victoria*. The location of the fusion protein can then be determined by fluorescence microscopy.

One of the most important advances reported recently has been the development of inducible expression systems for *T. brucei*, *Leishmania* and *T. cruzi* (Figure 3). Because trypanosomatids lack inducible promoters, use was made of the *Escherichia coli* tetracycline (Tc)-responsive Tet repressor (*tet*R) and the T7 promoter as the central components of the system. Trypanosomes were first transformed to constitutively express the Tet repressor and T7 RNA polymerase following integration of the genes into an appropriate site, such as the tubulin locus. An expression construct was then produced in which a reporter gene was placed under the control of the T7 promoter, with the Tet operator sequences placed immediately adjacent to the site of transcription initiation (Figure 3). In *T. brucei* best results have been achieved when this construct was targeted in the reverse orientation into the rDNA non-transcribed spacer. Addition of as little as 50 ng/ml Tc could produce a 100,000-fold increase in the activity of the luciferase reporter; when Tc was removed this fell to background levels. In *T. cruzi*, inducible expression can even be achieved when the reporter gene is maintained in an episomal construct (Wen et al., 2001). These inducible systems will be crucial to the study of genes that express products that are potentially toxic, or which function in cell-cycle control or differentiation.

Another significant area of progress has been in the application of RNA interference (RNAi) technology. RNAi is based on a gene-specific silencing mechanism that is widespread in eukaryotes and seems to have evolved as an anti-viral strategy and as a means of suppressing transposon activity (Sharp, 2001). The current model for RNAi envisages that double-stranded RNA is first processed to yield small fragments of 21-23 nucleotides. These fragments then bind to complementary regions of the target mRNA and identify it for enzyme-mediated cleavage. In *T. brucei*, RNAi has been achieved by expressing identical target fragments of DNA arranged in a "head-to head" orientation, separated by a "stuffer" region. Transcripts expressed in this context form a double stranded helix with a hairpin loop.

Depending on the specific mRNA species, this approach can lead to a >90% reduction in the level of the target transcript (Wang et al., 2000). The best results, in terms of inferring the function of down-regulated genes, have been achieved by inducible expression of RNAi (Figure 3). The application of RNAi to *T. cruzi* has not yet been reported, although several groups, including ourselves are addressing the problem. Given the time that it takes to carry out the two rounds of targeted gene deletion necessary to produce null mutants in *T. cruzi* (6-9 months), and the multicopy nature of many genes, RNAi could be particularly effective for high-throughput analysis of gene function in this parasite.

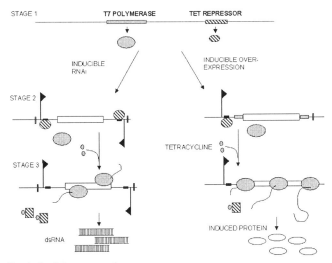

Figure 3 Tetracycline inducible systems for trypanosomes:
These systems are based on using the bacterial tetracycline repressor protein to block activity of a heterologous promoter sequence until the addition of tetracycline allows transcription to proceed. This system in protozoa was first developed in *T. brucei* by Wirtz and Clayton (1995) with subsequent modification by Wirtz et al., (1998)
Stage 1: A parasite line is created which expresses the T7 RNA polymerase and Tet repressor genes. This is the basic strain for inducible experiments.
Stage 2: The basic strain is transformed with one of the two constructs shown. For inducible protein expression the gene of interest (⬜) flanked by processing signals is cloned downstream of a tet repressible T7 promoter ▶). For RNAi the gene of interest is flanked on both sides by tet repressible T7 promoters in opposing orientations. This latter vector is based on pZJM (Wang et al., 2000).
Stage 3: The transformants are treated with tetracycline which binds to the repressor protein, causing it to dissociate from the DNA and allowing the T7 RNA polymerase to bind, thus inducing transcription. T7 transcription is limited to the gene of interest by the incorporation of T7 transcriptional terminators (▦).

As an additional approach to genome analysis we have developed the procedure of transfection-mediated chromosome dissection for *T. cruzi* (Figure 4). This technique enables large regions of a chromosome to be deleted in a single transfection experiment. It involves the use of a linear DNA vector that contains an array of telomeric hexamer repeats at one end and a targeting sequence at the other. Integration into the target locus by single-crossover results in a truncated chromosome, with a new telomere supplied by the vector. Using this technique we have been able to delete 700 kb of DNA in a single transfection. The availability of a complete genome sequence will allow the entire repertoire of genes deleted in each experiment to be identified. This will permit a more detailed analysis of these genes to be undertaken using a "single knockout" strategy. In addition, this approach may lead to the identification of regions containing dose-dependent genes, provide a means of delineating elements necessary for chromosome maintenance, contribute to our understanding of chromosome structure, and can also be used to confirm chromosome mapping data.

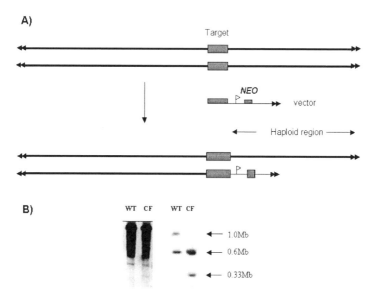

Figure 4. Transfection-mediated chromosome dissection in *T. cruzi*.
A) The strategy: After ligation of the target DNA into the vector, the construct is linearised and used to transfect *T. cruzi*. The flagged region in the vector represents the ribosomal DNA promoter. Diploid chromosomes containing the target sequence are shown. The vector is designed to integrate by single crossover to produce a truncated chromosome with a new telomere (indicated by filled triangles) supplied by the vector. By cloning the target DNA in the opposite orientation from that shown here it is possible to delete the opposite arm of the chromosome. The end result is to create partially monosomic mutants in which the genes in the "haploid" region are single copy in the genome.
B) The deletion of 700 kb from the 1000 kb homologue of chromosome III: We used the approach outlined to target the truncation of chromosome III. In the CL Brener clone of *T. cruzi* that is being used for the genome project, this chromosome exists as homologues of 600 and 1000 kb. Chromosomes from wild type (WT) and transfected (CF) parasites were separated by pulse field gel electrophoresis. A photograph of the ethidium bromide stained gel, overexposed to highlight the low molecular weight chromosomes, is shown (left). The gel was Southern blotted and probed with cruzipain, a gene located on chromosome III. The autoradiograph (right) shows the 600 and 1000 kb chromosome homologues in WT cells. In transfected cells, integration has resulted in a deletion of 700 kb from the larger homologue resulting in a truncated chromosome of 330 kb.

ANALYSIS OF GENE FUNCTION IN *T. CRUZI* USING TRANSFECTION

One of the first *T. cruzi* genes to be investigated at the level of function using transfection-based procedures was the surface glycoprotein, *gp72*. This example is worth considering in detail since it serves to illustrate the concept and methodology of the approach. Null mutants were successfully produced by consecutive deletion of both copies of *gp72* and their replacement with the *neo'* and *hyg'* selectable marker genes. At the morphological level, the null mutants were found to have an abnormal flagellar phenotype in which attachment to the parasite cell body was disrupted. Although they were still able to infect mammalian cells, the mutants had a significant impairment in their ability to survive within the triatomine vector. When the *gp72* gene was replaced using an episomal expression system the mutant phenotype could be complemented and the normal flagellar phenotype restored (Nozaki and Cross, 1994). This step is crucial when addressing gene function using targeted insertion/disruption. The ability to complement the mutant excludes the possibility that the observed phenotype could have arisen from an unexpected effect of the integration event on the expression of linked genes, rather than from deletion of the targeted allele.

The invasion of mammalian cells by *T. cruzi* is a complex process, but the mechanisms involved have been shown to be amenable to dissection by transfection studies. The role of the parasite serine hydrolase, oligopeptidase B, in cell invasion is one example of the application of this approach (Caler et al., 1998). Invasion of host cells by *T. cruzi* is thought to be dependent on the ability of the parasite to induce the recruitment and fusion of lysosomes at the site of entry. Evidence suggests that this involves the parasite-induced triggering of a transient increase in the concentration of free intracellular calcium ions in the host. It has been suggested that oligopeptidase B is a processing enzyme and that its role is to generate an active Ca^{2+} agonist from a *T. cruzi* cytosolic precursor. Targeted deletion of the oligopeptidase B gene resulted in parasites that were defective in their ability to invade mammalian cells or to establish an infection in mice. Furthermore, the observation that addition of recombinant oligopeptidase B to extracts of null mutant cells could restore Ca^{2+} signalling activity, confirmed the role of this enzyme in the generation of the agonist from a parasite precursor.

Other determinants of virulence that have been subjected to functional analysis using transfection procedures include Tc52, a protein with thioltransferase activity related to the thioredoxin/glutaredoxin family (Allaoui et al., 1999). Interestingly this protein has also been reported to have a role in the immunosuppression associated with Chagas disease. Isolation of *Tc52* null mutants was found not to be achievable suggesting that the protein is essential for viability. Parasites containing a single disrupted allele displayed no obvious growth defect when cultured as epimastigotes. However, the ability of these cells to develop into metacyclic trypomastigotes, or to proliferate *in vitro* or *in vivo,* was considerably reduced. Confirmation that this was due to decreased levels of Tc52 was established using an episomal vector containing the *Tc52* gene to complement the mutant phenotype.

The function of Tc52 has also been explored by producing transformed cells that were manipulated to express high levels of *Tc52* antisense RNA. Using this approach it was possible to reduce the level of expressed protein to about 10% of normal levels. These cells displayed a mutant phenotype that was similar to that of the single knockouts in terms of intracellular growth and differentiation. A parallel approach has also been taken to investigate the biological role of trypanothione reductase (TR). This enzyme is thought to be central to thiol metabolism in *T. cruzi*, and other trypanosomatids, and functions by maintaining the parasite-specific thiol trypanothione in its reduced form. Attempts were made to down-regulate TR activity in *T. cruzi* by expression of antisense RNA from a plasmid construct (Tovar and Fairlamb, 1996). However, it was not possible in these experiments to produce genetically modified parasites that had reduced levels of TR activity. In all transformants that were analysed the input plasmid was found to have undergone rearrangements that prevented expression of the antisense transcript. This could indicate that down-regulation of TR activity is detrimental to *T. cruzi* and that under culture conditions, only those cells that contain a defective, rearranged plasmid are able to be selected. Consistent with this interpretation, gene deletion experiments have shown that *TR* is an essential gene in *Leishmania*. With *T. brucei,* null mutants are avirulent and have increased sensitivity to oxidative stress.

THE *T. CRUZI* GENOME PROJECT AND "POST-GENOME" TECHNOLOGIES

The *T. cruzi* genome initiative was launched in 1994 as part of a WHO-sponsored programme of parasite genome analysis (The *Trypanosoma cruzi* genome consortium, 1997). The CL-Brener clone, which has a haploid genome content of approximately 45 Mbp, was selected for sequencing. Initial studies focussed on karyotype analysis, mapping and sequencing of chromosome III, and the sequencing of expressed sequence tags (ESTs). More than 10,000 *T. cruzi* EST sequences are now available (http://web.genpat.uu.se/TcruziESTc/index.html). Since 2000, a consortium of laboratories from The Institute for Genomic Research (TIGR) and the Seattle Biomedical Research Institute in the USA, and from Uppsala University in Sweden have begun the large scale sequencing component of the project. The first phase of this has involved the determination of 48,000 end sequences of clones from bacterial artificial chromosome (BAC) libraries (average insert size ~100 kb). A fully annotated genome sequence is expected in 2004-2005.

The genome project is having a major impact on the research strategies applied to *T. cruzi*. The focus is changing from the characterisation of individual determinants to a more comprehensive analysis of function at a genome-wide level, a multidisciplinary approach that has been termed "functional genomics". Several developing technologies are helping to fuel this new integrated strategy. Broadly speaking these can be divided into the new high throughput methodologies for gene knockout/disruption and the new techniques for profiling gene expression at a global level (microarrays and proteomics).

As outlined above, RNAi techniques offer a rapid approach for determining phenotypes associated with the disruption of gene function. This procedure avoids the laborious process of attempting to produce null mutants by two rounds of targeted gene deletion. Similarly, chromosome dissection techniques (Figure 4) have the potential to be used for the creation a bank of partially monosomic mutants that covers a large percentage of the parasite genome. This should allow non-truncated chromosome homologues to be more rapidly screened for genes that are essential or non-essential using a "single knockout" strategy

For expression studies, microarray analysis can provide a snapshot of the relative expression levels of thousands of genes simultaneously (the transcriptome) (for review, see Lockhart and Winzeler, 2000). The technique involves spotting DNA by robot onto glass slides in high density arrays, up to 10,000 samples per cm^2. For smaller bacterial genomes, it is possible to obtain coverage of the entire genome on a single slide. With higher organisms chromosome-specific arrays are more common. DNA fragments can be arrayed in the form of oligonucleotides, cDNAs or PCR products. The most common use of microarrays has been to examine differential expression of mRNAs, such as at different life-cycle stages, or after drug treatment. Typically cDNA labelled with different fluorescent dyes will be prepared from the two mRNA populations. After hybridisation the array can be scanned at the wavelength optima of the different dyes to determine the relative abundance of each transcript under the two different sets of conditions. It has been argued however, that microarray analysis may be less useful when applied to trypanosomatids because mRNA levels often do not reflect changes in the abundance of the corresponding protein. This reflects that control of gene expression in these parasites often operates at the translational/post-translational level. In the case of *T. cruzi,* interpretation of the data may also be complicated by the high number of related genes.

As an alternative or complementary approach to expression profiling the proteomics route offers many advantages (for review, see Pandey and Mann, 2000). The power of the technique results from ability to relate multiple protein spots on a 2-D polyacrylamide gel to the genome sequence database using mass spectrometry. In a process that can be highly automated, protein spots are excised, fragmented by enzyme digestion and peptide mass-fingerprints obtained. Individual proteins can then be identified with a success rate that depends largely on the completeness of the database. The extent and nature of protein modifications, including phosphorylation and glycosylation, are other features that can be addressed by these approaches. Proteome analysis is a rapidly developing area and technical advances, particularly in field of mass spectrometry, offer the promise of even greater analytical resolution. Applications to *T. cruzi* include the profiling of stage-specific expression, identification of proteins induced by environmental stimuli or drug treatment, and the dissection of signalling pathways. In addition the ability to incorporate specific *T. cruzi* null mutants into a proteome study offers the possibility of defining the biochemical phenotype in very precise terms.

In summary, research on *T. cruzi* is undergoing a revolution. The progress of the genome project, the development of increasingly flexible and sophisticated genetic tools, and the availability of new techniques for profiling gene expression will provide an integrated framework to dissect the complex biology of this intractable parasite. It will be of particular importance to identify new targets for chemotherapy, to identify genes that have a central role in parasite development and virulence, and to gain a greater understanding of disease pathogenesis.

References

Aguero F, Verdun RE, Frasch AC, Sanchez DO. 2000. A random sequencing approach for the analysis of the *Trypanosoma cruzi* genome: general structure, large gene and repetitive DNA families, and gene discovery. Genome Res 10:1996-2005.

Allaoui A, Francois C, Zemzoumi K, Guilvard E, Ouaissi A. 1999. Intracellular growth and metacyclogenesis defects in *Trypanosoma cruzi* carrying a targeted deletion of a Tc52 protein-encoding allele. Mol Microbiol 32:1273-1286.

Caler EV, Vaena de Avalos S, Haynes PA, Andrews NW, Burleigh BA. 1998. Oligopeptidase B-dependent signaling mediates host cell invasion by *Trypanosoma cruzi*. EMBO J 17:4975-4986.

Clayton CE. 1999. Genetic manipulation of kinetoplastida. Parasitol Today 15:372-378.

D'Orso I, Frasch AC. 2001. Functionally different AU- and G- rich *cis*-elements confer developmentally regulated mRNA stability in *Trypanosoma cruzi* by interaction with specific RNA-binding proteins. J Biol Chem 276:15783-15793.

Kelly JM. 1995. Trypanosomatid shuttle vectors: new tools for the functional dissection of parasite genomes. Parasitol Today 11:447-451.

Kelly JM. 1997. Genetic transformation of parasitic protozoa. Adv Parasitol 39:227-270.

Lockhart DJ, Winzeler EA. 2000. Genomics, gene expression and DNA arrays. Nature 405:827-836.

Machado CA, Ayala FJ. 2001. Nucleotide sequences provide evidence of genetic exchange among distantly related lineages of *Trypanosoma cruzi*. Proc Natl Acad Sci USA 98:7396-7401.

Martinez-Calvillo S, Lopez I, Hernandez R. 1997. pRIBOTEX expression vector: a pTEX derivative for rapid selection of *Trypanosoma cruzi* transfectants. Gene 199:71-76.

Miles AA, Yeo M, Gaunt M. 2002. Genetic diversity of *Trypanosoma cruzi* and the epidemiology of Chagas disease. In: Molecular Mechanisms in the Pathogenesis of Chagas Disease. Kelly JM (ed.) Eurekah.com/Landes Bioscience, Austin, Texas (in press).

Nozaki T, Cross GAM. 1994. Functional complementation of glycoprotein 72 in a *Trypanosoma cruzi* glycoprotein 72 null mutant. Mol Biochem Parasitol 67:91-102.

Pandey A, Mann M. 2000. Proteomics to study genes and genomes. Nature 405:837-846.

Schnaufer A, Panigrahi AK, Panicucci B, Igo RP, Salavati R, Stuart K. 2001. An RNA ligase essential for RNA editing and survival of the bloodstream form of *Trypanosoma brucei*. Science 291:2159-2162.

Sharp PA. 2001. RNA interference – 2001. Genes Dev 15:485-490.

Simpson L, Thiemann OH, Savill NJ, Alfonzo JD, Maslov AD. 2000. Evolution of RNA editing in trypanosome mitochondria. Proc Natl Acad Sci USA 97:6986-6993.

Stothard JR, Frame IA, Miles MA. 1999. Genetic diversity and genetic exchange in *Trypanosoma cruzi*: dual drug-resistant "progeny" from episomal transformants. Mem Inst Oswaldo Cruz 94: (Suppl. 1):189-193.

The *Trypanosoma cruzi* genome consortium. 1997. The *Trypanosoma cruzi* genome initiative. Parasitol Today 13:16-22.

Tibbetts RS, Klein KG, Engman DM. 1995. A rapid method for protein localization in trypanosomes. Exp Parasitol 80:572-574.

Tomas AM, Kelly JM. 1996. Stage-regulated expression of cruzipain, the major cysteine protease of *Trypanosoma cruzi* is independent of the level of RNA. Mol Biochem Parasitol 76:91-103.

Tovar J, Fairlamb AH. 1996. Extrachromosomal, homologous expression of trypanothione reductase and its complementary mRNA in *Trypanosoma cruzi*. Nucleic Acids Res 24:2942-2949.

Wang Z, Morris JC, Drew ME, Englund PT. 2000. Inhibition of *Trypanosoma brucei* gene expression by RNA interference using an integratable vector with opposing T7 promoters. J Biol Chem 275:40174-40179.

Wen LM, Xu P, Benegal G, Carvaho MR, Butler DR, Buck GA. 2001. *Trypanosoma cruzi*: exogenously regulated gene expression. Exp Parasitol 97:196-204.

Wirtz E, Clayton C. 1995. Inducible gene expression in trypanosomes mediated by a prokaryotic repressor. Science 268:1179-1183.

Wirtz E, Hoek M, Cross GA. 1998. Regulated processive transcription of chromatin by T7 RNA polymerase in *Trypanosoma brucei*. Nucleic Acids Res 26:4626-34.

TRYPANOSOMA CRUZI CELL INVASION MECHANISMS

N. Yoshida
Departamento de Microbiologia, Imunologia e Parasitologia, Escola Paulista de Medicina, Universidade Federal de São Paulo, R. Botucatu, 862, São Paulo, S.P., Brasil

ABSTRACT

Mammalian cell invasion by *Trypanosoma cruzi* is a multi-step process that requires the interaction of various parasite and host cell molecules and the activation of signal transduction pathways in both cells. Depending on the parasite strain, developmental stage and the type of target cell, distinct sets of molecules may be engaged. The metacyclic trypomastigote, which initiates infection of the mammalian host, expresses stage-specific surface glycoprotein gp82 and the mucin-like molecule gp35/50. The binding of these proteins to the target cell during infection triggers Ca^{2+} responses in both cells. Cell-derived trypomastigotes trigger signaling pathways from receptors for agonists, such as bradykinin and TGFβ, and from receptors for a trypomastigote Ca^{2+}-agonist, generated from a precursor molecule by oligopeptidase B. Trypomastigote surface molecules, such as trans-sialidase, the stage-specific mucin SSp-3 and Tc85, which contains binding sites for laminin and cytokeratin 18, may also be involved in cell invasion. In host cells, Ca^{2+} released from internal compartments in IP_3-dependent manner is required for translocation of lysosomes to the site of trypomastigote entry, a process that contributes to the formation of the vacuole membrane surrounding the parasite during penetration.

INTRODUCTION

To infect mammalian hosts, *Trypanosoma cruzi* must invade cells, replicate intracellularly and, upon cell rupture, disseminate to diverse organs and tissues. These additional rounds of the intracellular cycle take place until parasite proliferation is controlled by the host immune response. In natural infection, the contact of *T. cruzi* with host cells that results in invasion is established by metacyclic trypomastigotes, which are released in the feces of triatomine insects during blood meals. The metacyclic forms can invade and survive inside professional phagocytes as well as in nonphagocytic cells. The replicative epimastigotes from the vector may also gain access to the host, but are unable to enter nonphagocytic cells and are destroyed within macrophages. Host cell penetration by metacyclic trypomastigotes proceeds with the parasite internalized in a membrane-bounded vacuole. Several hours after invasion, metacyclic forms escape from the parasitophorous vacuole to the cytoplasm where they transform into amastigotes. Following replication as amastigotes by binary fission, *T. cruzi* differentiates into trypomastigotes that are released in the bloodstream upon host cell lysis. These blood trypomastigotes, endowed with the capacity to invade a variety of cell types, are responsible for propagating the infection.

Most studies on cell invasion by *T. cruzi* have been performed with cultured counterparts of metacyclic and bloodstream trypomastigotes, and to a lesser extent with amastigotes released from infected cells or generated extracellularly. A variety of nonphagocytic mammalian cell lines have been used as targets for invasion, in addition to macrophages - the professional phagocytes that presumably play a role *in vivo* in the establishment of *T. cruzi* infection. The picture emerging from these studies is complex, for it appears that the mechanism of invasion may vary depending on the parasite form and the type of host cell under analysis. Different developmental stages engage distinct sets of molecules to interact with particular cell types and activate specific signal transduction pathways. In addition, strain-specific differences may further contribute to the differences in parasite-host cell interaction: *in vivo* and *in vitro* experiments suggest that there are marked differences in infectivity between *T. cruzi* strains. Notwithstanding all this complexity, a unifying theme is that mammalian cell invasion by *T. cruzi* trypomastigotes is a multi-step process involving various parasite and host cell molecules. Further, that it is accomplished, through a concerted series of events, leading to intracellular Ca^{2+} mobilization in both trypanosome and host cells and to lysosome recruitment at the site of parasite entry, in nonphagocytic cells.

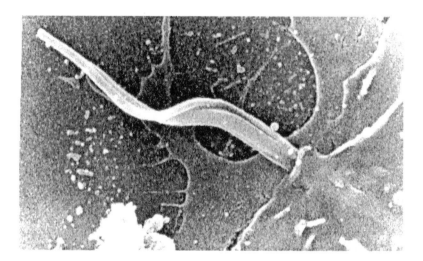

Figure 1. Scanning electron micrograph of *T. cruzi* trypomastigote entering nonphagocytic mammalian cell. (Courtesy of Sergio Schenkman).

Signaling in *T. cruzi* and host cell

A number of studies have shown that internalization of *T. cruzi* trypomastigotes (Figure 1) requires the activation of signal transduction mechanisms, leading to an increase in intracellular Ca^{2+} concentration ($[Ca^{2+}]_i$) in both the parasite and the host cell (Moreno et al., 1994, Burleigh & Andrews, 1998). Thus, interfering with intracellular Ca^{2+} mobilization, either by pretreatment of *T. cruzi* or the host cell with Ca^{2+} chelators or agents that deplete $[Ca^{2+}]_i$, reduces parasite entry. As Ca^{2+} signal is triggered in distinct mammalian cell types from different species (Table 1) either by metacyclic or tissue culture trypomastigotes, but not by noninfective epimastigotes, stage-specific components must be involved.

The stage-specific surface glycoprotein gp82, expressed in metacyclic trypomastigotes of different *T. cruzi* strains, binds to nonphagocytic mammalian cells in a receptor-mediated manner and triggers Ca^{2+} response in both cells. Evidence from experiments with the highly invasive *T. cruzi* CL strain suggests that the interaction of gp82 with host cells induces a signaling cascade in the parasite. This leads to sequential activation of protein tyrosine kinase (PTK) and phospholipase C, with generation of inositol 1,4,5-triphosphate (IP$_3$), which promotes Ca^{2+} mobilization from IP$_3$-sensitive stores. PTK activation, Ca^{2+} response and parasite internalization are inhibited by recombinant gp82, as well as by pretreatment of metacyclic trypomastigotes with the PTK inhibitor genistein. Ca^{2+} signaling and the ability to invade epithelial cells are reduced in parasites treated with phospholipase C inhibitor U73122, or with drugs such as heparin and thapsigargin that affect Ca^{2+} release by competitively blocking IP$_3$ receptor or by depleting intracellular Ca^{2+} stores. In the same manner as the metacyclic forms, the tissue culture trypomastigotes have their infectivity reduced upon treatment with PTK inhibitor, but in this case the parasite molecule that triggers the PTK activation is not known. Another metacyclic trypomastigote cell adhesion molecule with signaling activity is the mucin-like surface glycoprotein gp35/50. Monoclonal antibodies directed to the carbohydrate moiety of this molecule induce a Ca^{2+} response in the parasites. Both this mucin molecule and gp82 are anchored to the plasma membrane via glycosylphosphatidylinositol (GPI) and therefore associate only with the outer leaflet of the lipid bilayer. To transduce the external signals to the parasite interior, these GPI-anchored glycoproteins probably interact with plasma membrane molecules asscociated with the downstream components of the signaling cascade.

Table 1. Mammalian cells used in assays of *T. cruzi* invasion and Ca^{2+} signaling.

Cell type	*T. cruzi*-induced Ca^{2+} response	Infection by *T. cruzi*
Chinese hamster ovary (CHO) cell	+	Susceptible
Human carcionoma-derived epithelial HeLa cell	+	Susceptible
Human leukemic K562 cell	–	Resistant
Human umbilical vein endothelial cell (HUVEC)	+	Susceptible
L6E9 myoblast	+	Susceptible
Madin-Darby canine kidney (MDCK) cell	+	Susceptible
Normal rat kidney (NRK) fibroblast	+	Susceptible
Primary canine cardiac myocoyte	+	Susceptible
Vero cell derived from African green monkey fibroblast	+	Susceptible

In target cells, signal transduction pathways may be activated by *T. cruzi* molecules, which are either secreted or expressed on the parasite surface. The metacyclic stage gp82 induces Ca^{2+} mobilization in epithelial cells and this response can be inhibited by 3F6; a monoclonal antibody directed to a peptide sequence of gp82 and capable of decreasing parasite internalization. Ca^{2+} response is also triggered in epithelial cells by metacyclic trypomastigote gp35/50. Upon interaction with diverse cell types, a factor that triggers Ca^{2+} response is generated in tissue culture trypomastigotes from a precursor molecule through the action of a cytosolic enzyme, oligopeptidase B (Burleigh and Andrews, 1998). The oligopeptidase B null mutant trypomastigotes are defective in mobilizing Ca^{2+} from thapsigargin-sensitive stores in mammalian cells and have diminished invasive capacity (Caler et al., 1998). Purified oligopeptidase B alone is unable to induce a Ca^{2+} response in the host cell. Based on experimental evidence, it has been proposed that the Ca^{2+} agonist generated by oligopeptidase B is exported from the parasite, and binds to a receptor on the surface of target cells. This activates phospholipase C and generates IP_3; IP_3 then binds to its receptor on the membrane of the endoplasmic reticulum and promotes Ca^{2+} release (Figure 2).

The host cell receptors for metacyclic stage gp82 or gp35/50, or tissue culture trypomastigote Ca^{2+}-agonist, have not been characterized. In nonphagocytic cells, they are not associated with PTK. Treatment of different target cells, such as NRK fibroblasts and HeLa epithelial cells, with PTK inhibitor genistein does not affect trypomastigote invasion. On the other hand, trypomastigote entry into macrophages, a process that is also dependent on $[Ca^{2+}]_i$ increase, requires PTK activity (reviewed in Burleigh and Andrews, 1998). *T. cruzi* may also signal through mammalian cell receptors for physiologically relevant ligands. In a study using tissue culture trypomastigotes from a clone of *T. cruzi* strain Dm28, the invasion of HUVECs or CHO cells overexpressing the B_2 type of bradykinin receptor (CHO-B_2R) was shown to be subtly modulated by the combined activities of kininogen, kininogenases and kinin-degrading peptidases (Scharfstein et al., 2000). Captopril, an inhibitor of bradykinin degradation by kininase II, potentiated the trypomastigote entry into HUVECs and CHO-B_2R, but not into mock-transfected CHO cells, whereas the B_2R antagonist HOE 140, or monoclonal antibody to bradykinin, blocked these effects. Cruzipain, a *T. cruzi* serine proteinase, seems to play a role in generating the kinin agonist from cell-bound kininogen (Figure 2). The purified enzyme enhanced parasite invasion and triggered Ca^{2+} mobilization in CHO-B_2R, this Ca^{2+} response being abrogated by HOE 140 or cruzipain inhibitor E-64. As the kinin-mediated signal transduction route is not activated in Vero cells or L6E9 myoblasts, and its use is not ubiquitous in *T. cruzi* strains (Scharfstein et al., 2000), it may operate only in some specific circumstances.

In another study, the transforming growth factor-β (TGFβ) pathway was shown to be required for *T. cruzi* invasion of epithelial cells (Ming et al., 1995). Trypomastigotes attached to TGFβ receptor-deficient epithelial cell lines, but failed to penetrate. Susceptibility to *T. cruzi* infection was restored by transfection with TGFβ receptor genes. Since trypomastigotes release a factor that induces a TGFβ responsive reporter gene in TGFβ-sensitive cell lines, it has been suggested that a *T. cruzi* TGFβ-like ligand activates the mammalian cell TGFβ signaling pathways and promotes parasite invasion. In macrophages, the heterodimeric β1-integrins, which belong to a ubiquitous family of integral membrane proteins that link the extracellular matrix to the cortical cytoskeleton, may be involved in signal transduction and *T. cruzi* internalization. A monoclonal antibody specific for β1-integrins and effective in inhibiting trypomastigote attachment and the binding of fibronectin to β1-integrin interfered with invasion, whereas other antibodies directed to different epitopes and not involved in binding fibronectin did not (reviewed in Burleigh and Andrews, 1995).

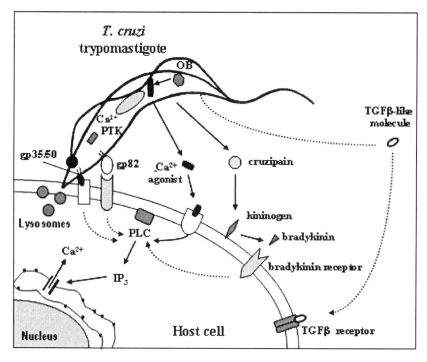

Figure 2. Schematic representation of signaling molecules and pathways that may be activated during *T. cruzi* entry into nonphagocytic mammalian cells. Interaction of metacyclic trypomastigotes through the stage-specific surface glycoprotein gp82 with its receptor induces the activation of parasite protein tyrosine kinase (PTK) that leads to Ca^{2+} mobilization. In the target cell, gp82 triggers a signal transduction pathway that involves Ca^{2+} release, possibly from endoplasmic reticulum in a manner dependent of IP_3 generated by phospholipase C (PLC). In addition to gp82, metacyclic forms may engage the mucin-like surface molecule gp35/50 to trigger Ca^{2+} signaling in both cells. Conversely, tissue culture trypomastigotes secrete a Ca^{2+} agonist derived from a precursor molecule by the action of a cytosolic oligopeptidase B (OB), which binds to its receptor on the host cell surface and triggers the Ca^{2+} mobilization route involving PLC and IP_3. Depending on the parasite-target cell pair, cell-derived trypomastigotes may signal through host cell receptors for physiologically relevant agonists such as bradykinin and transforming growth factor-β(TGFβ). In the bradykinin-mediated pathway, the secreted form of *T. cruzi* proteinase cruzipain would generate the agonist that signals through the bradykinin receptor leading to Ca^{2+} mobilization by acting on the host kininogen. Signaling through TGFβ receptors involves a postulated TGFβ-like molecule released by trypomastigotes. The Ca^{2+} response induced in the host cell promotes lysosome translocation to the site of parasite entry. Solid arrows, supported by experimental evidences. Dashed arrows, postulated events.

Figure 2 depicts schematically the diverse signaling molecules and pathways that may be involved in the entry of *T. cruzi* metacyclic or cell-derived trypomastigotes into nonphagocytic mammalian cells. Following ligand binding, the different host cell receptors may associate with distinct molecules on the cytoplasmic side of the plasma membrane. The result of these associations is the activation of PLC and IP_3-mediated Ca^{2+} efflux from internal compartments.

Lysosome recruitment

The Ca^{2+} signal triggered in host cells by contact with trypomastigotes or parasite soluble factors is thought to contribute to lyosome translocation, which is required for *T. cruzi* entry into nonphagocytic cells. Before the onset of internalization, lysosomes cluster at the site of trypomastigote attachment, in close proximity to the host cell plasma membrane (reviewed in Burleigh and Andrews, 1995). Lysosomal markers can be detected in partially or fully formed intracellular vacuoles surrounding parasites. Several lines of experimental evidence indicate that the fusion of lysosomes with the plasma membrane promotes *T. cruzi* internalization, by providing the membrane required for vacuole formation. Acidification of the cytosol induces migration of lysosomes from the perinuclear region to the cell periphery and enhances trypomastigote entry, whereas the opposite effect is observed by perinuclear clustering of lysosomes induced by alkalinization. In addition, *T. cruzi* invasion is reduced when lysosome fusion capacity is diminished, for instance by loading lysosomes with sucrose. It has been shown that the Ca^{2+}-dependent exocytosis of lysosomes in NRK cells is potentiated by intracellular cAMP, whose levels are elevated upon interaction with trypomastigotes but not with noninfective epimastigotes (Rodriguez et al., 1999). This second messenger is synthesized from ATP by adenylyl cyclase, which is activated upon interaction with the α-subunit of stimulatory G protein. Activation of adenylyl cyclase through cell stimulation with α-adrenergic agonist isopreterenol enhances the Ca^{2+}-dependent exocytosis of lysosomes and *T. cruzi* invasion, whereas the adenylyl cyclase inhibitors have the opposite effect. More recently, Caler et al. (2001) have shown that sinaptotagmin VII, a ubiquitously expressed sinaptotagmin isoform localized on the membrane of lyososomes in different cell types, regulates exocytosis of these organelles and mediates *T. cruzi* invasion. A marked inhibition of *T. cruzi* entry was observed in CHO cells loaded with antibodies directed to the Ca^{2+}-binding C_2A domain of sinaptotagmin VII, which effectively inhibit the Ca^{2+}-triggered exocytosis of lysosomes when added to permeabilized cells. Ca^{2+}-dependent exocytosis of lysosomes was also inhibited by the soluble recombinant construct containing the C_2A domain of sinaptotagmin VII, but not by the construct containing the C_2A domain of sinaptotagmin I, the exclusively neuronal isoform present on the membrane of synaptic vesicles.

For the contact and fusion of lysosomes with the plasma membrane, the actin cytoskeleton must be rearranged. Consistent with this, treatment of cells with cytochalasin D, a drug that disrupts microfilaments and prevents the formation of plasma membrane extensions mediated by actin, enhances Ca^{2+}-dependent exocytosis of lysosomes and facilitates *T. cruzi* invasion. The preference of *T. cruzi* trypomastigotes to enter polarized cells through their basolateral domains and flat lamellipodia close to the cell margins, where assembly and disassembly of the cortical actin cytoskeleton occurs more dynamically, is also compatible with the mechanism of *T. cruzi* entry independent of actin polymerization. Unlike trypomastigotes, amastigotes attach to the dorsal surface of cells and their internalization is greatly inhibited by disruption of host cell microfilaments with cytochalasin D. (reviewed in Burleigh and Andrews, 1998).

Mucins, sialic acid and trans-sialidase

Among the *T. cruzi* surface glycoproteins that have been implicated in host cell invasion are the mucin-like molecules, a group of highly glycosylated GPI-anchored proteins rich in threonine, serine and proline residues. These are expressed in high copy numbers in both insect and mammalian stages of the parasite (Acosta-Serrano et al., 2001). *T. cruzi* mucins are encoded by a multigene family and contain a unique type of glycosylation consisting of several sialylated O-glycans, linked to the protein backbone via N-acetylglucosoamine residues. They constitute the major substrate for the *T. cruzi* trans-sialidase (TS), the enzyme that transfers $\alpha(2\text{-}3)$-linked sialic acid from extrinsic host-derived macromolecules to parasite surface acceptor molecules containing β-galactose (Schenkman et al., 1994).

In metacyclic trypomastigotes, the prevalent mucin is gp35/50. Based on reactivity with specific monoclonal antibodies, two variant forms of gp35/50 have been identified. They are differentially expressed in *T. cruzi* strains that differ in infectivity *in vivo* and *in vitro* (Ruiz et al., 1998). One variant, recognized by the monoclonal antibody 10D8 (which is directed to an epitope containing β-galactose furanose), was shown to bind to nonphagocytic cells in a receptor-dependent manner and to induce Ca^{2+} mobilization. It is present in less invasive strains and appears to participate in parasite internalization. Metacyclic trypomastigotes of 10D8-reactive *T. cruzi* strains have their entry into nonphagocytic cells reduced either by the antibody or the purified gp35/50. The sialic acid residues of this variant form of mucin may impair invasion. Removal of sialic acid from gp35/50 purified from the poorly invasive G strain, by treatment with sialidase augments its Ca^{2+} signaling activity towards target cells and, consistent with this, desialylation of G strain metacyclic forms results in increased parasite infectivity, this effect being reversed by mucin resialylation. The role played in cell invasion by the other variant of gp35/50, undetectable by monoclonal antibody 10D8 and expressed in highly invasive *T. cruzi* strains, such as Y and CL, is not clear.

In trypomastigotes derived from infected mammalian cells, the mucins were described as a group of molecules ranging from 60 to 200 kDa, initially identified as the stage-specific antigen Ssp-3. Shortly after release from host cells, trypomastigotes contain little or no sialic acid on their surface. Upon contact with mammalian cell membranes, extracellular matrix components or serum glycoproteins, the parasite acquires sialic acid, which is incorporated by TS mainly onto O-linked oligosaccharides of Ssp-3. As the infection of nonphagocytic cells by trypomastigotes is inhibited by monoclonal antibodies to a sialic acid-containing epitope generated by trans-sialidase, Ssp-3 may be involved (Schenkman and Eichinger, 1993), though the degree of mucin sialylation has little or no effect (Acosta-Serrano et al., 2001). Conversely, internalization of *T. cruzi* by macrophages is enhanced after desialylation of trypomastigotes and exposure of terminal _β-galactosyl residues. As D-galactose and N-acetyl-D-galactosamine inhibit infection of macrophages by trypomastigotes, the D-galactose/N-acetyl-galactosamine receptors could participate in the parasite recognition process (reviewed in Schenkman et al., 1994).

The degree of sialylation of *T. cruzi* seems to be irrelevant for cell invasion, whereas the amount of sialic acid on the host cell appears to positively regulate invasion of metacyclic and tissue culture trypomastigotes of different *T. cruzi* strains (reviewed in Schenkman et al., 1994). Sialic acid-deficient mutants of CHO cells, Lec2, are poorly invaded by trypomastigotes as compared with the wild type cells, and invasion is restored after Lec2 resialylation. Also, the rate of infection by blood trypomastigotes is reduced after treatment of mouse fibroblasts with sialidase (Schenkman et al., 1994). In addition to its effect on invasion, host cell sialic acid may also influence escape of *T. cruzi* from the membrane vacuole that surrounds it. The luminal side of this vacuole is covered with lysosomal glycoproteins (lgps) containing sialic acid. Removal of sialic acid from lgps, which are substrates for *T. cruzi* TS, would facilitate the rupture of the vacuolar membrane mediated by TcTox. This pore-forming protein secreted by the parasite is immunologically related to C9, the terminal component of the complement system that polymerizes to form pores in plasma membranes. After desialylation, membranes are more susceptible to lysis with TcTox, which is active in an acidic environment required for escape. Compatible with this is the finding that trypomastigotes escape faster from vacuoles of the sialic acid deficient Lec2 cells than from vacuoles of parental cell lines (reviewed in Schenkman et al., 1994; Burleigh and Andrews, 1995).

What is the role played by *T. cruzi* TS in host cell invasion? Metacyclic trypomastigotes express an enzyme much less active than that of tissue culture trypomastigotes. Intracellular amastigotes lack TS, but the enzymatic activity reappears after amastigotes transform into trypomastigotes before they burst out of infected cells. TS continues to be expressed by the free extracellular trypomastigotes and large amounts of enzyme are also released into the culture medium (Schenkman et al., 1994). Experiments with *T. cruzi* strains Y, CL and G suggest that trypomastigotes do not depend on TS activity to enter host cells. The TS of trypomastigotes may have a lectin-like activity towards the fully sialylated glycoconjugates, like those generally present in host tissues. Thus, rather than removing the host cell sialic acid, it is possible that the enzyme binds to it (Figure 3) as an important step in the entry process. This is consistent with the finding that, of the large amounts of trans-sialidase in tissue culture trypomastigotes, some are devoid of hydrolytic or transfer activity but have acceptor-binding capability. Hence the proposal that the mammalian cell surface receptors for *T. cruzi* are sialylated molecules, where removal of the sialic acid by a sialidase would preclude invasion. A proposal further supported by the observation that the sialic acid-deficient CHO cells, Lec2, are invaded at lower levels than parental cells (Schenkman & Eichinger, 1993; Schenkman et al., 1994). On the other hand, working with *T. cruzi* clone Silvio X-10/4, Pereira et al. (1996) observed that only 20-30% of trypomastigotes released from infected cells expressed TS. These TS[+] trypomastigotes were highly invasive whereas TS[−] parasites, which represented the majority of the population, were inefficient in invading epithelial cells and fibroblasts. Addition of small amounts of exogenous trans-sialidase to suspensions of non-penetrating TS[−] parasites converted them to the highly invasive phenotype. How this is accomplished is not known. One possibility yet to be investigated is that the soluble TS triggers a host cell signaling cascade that promotes parasite invasion.

T. cruzi proteins that bind to extracellular matrix components

Proteins with adhesive properties to components of the extracellular matrix, identified in tissue culture trypomastigotes, have also been implicated in host cell invasion. One of these proteins is a 60 kDa trypomastigote-specific protein, designated penetrin, that binds to heparin, heparan sulfate and collagen. These extracellular matrix components inhibit the binding of penetrin to Vero cells and invasion of these cells by trypomastigotes (reviewed in Burleigh and Andrews, 1995). Another molecule is a trypomastigote 85 kDa protein that binds fibronectin and may use use this component to adhere to target cell membranes. Antibodies to fibronectin bind to parasites in a specific saturable manner and inhibit invasion. Soluble human fibronectin has an opposite effect, but peptides that mimic the fibronectin attachment domain competititively inhibit trypomastigote invasion (reviewed in Burleigh and Andrews, 1995). The laminin-binding Tc85-11, a member of surface glycoproteins known collectively as Tc85, may also play a role in cell invasion. This glycoprotein belongs to the gp85/trans-sialidase superfamily and is recognized by a monoclonal antibody that inhibits host cell invasion (Magdesian et al., 2001). Tc85-11 binds to isolated laminin and to entire cells through its carboxy-terminal segment. Experiments using synthetic peptides showed that the common sequence VTVXNVFLYNR of the gp85/ trans-sialidase supergene family is a mammalian cell-binding motif and that cytokeratin 18 (CK18) is the host cell receptor for this conserved motif (Magdesian et al., 2001). The metacyclic stage gp82, which is also a member of gp85/ trans-sialidase superfamily, lacks the ability to bind laminin, fibronectin or heparan sulfate. It contains the conserved motif VTVXNVFLYNR at the carboxy-terminus of the molecule, very close to the GPI anchor. A series of studies has shown that the cell binding site of gp82 is localized in its central domain (in partial overlap with the epitope for 3F6, a monoclonal antibody that inhibits parasite invasion), possibly formed by juxtaposition of two noncontiguous amino acid sequences separated by a hydrophobic segment.

Figure 3 shows schematically the various *T. cruzi* molecules, with adhesive properties towards host cell receptors and/or extracellular matrix components, which may play a role in the process of trypomastigote entry into nonphagocytic mammalian cells.

Concluding remarks

Several molecules of the infective trypomastigote stage of *T. cruzi* have been implicated in mammalian cell invasion. They are either secreted or expressed on the parasite surface and have the ability to adhere to host cells and/or to extracellular matrix components. Some of the parasite ligands have been shown to induce Ca^{2+} mobilization, which is required for lysosome translocation to the site of parasite entry. In the multi-step cell invasion process by *T. cruzi*, the different parasite molecules may play specific roles either in the initial attachment/signaling phase or in later internalization steps. The repertoires of parasite molecules that are engaged, may differ with the *T. cruzi* strain, infective developmental stage (metacyclic or cell-derived trypomastigote), and the type of target cell. The availability of multiple alternative routes to enter host cells may explain the capacity of *T. cruzi* to infect a wide variety of cell types and why complete inhibition of invasion by blocking a particular route is never achieved.

78 *Yoshida*

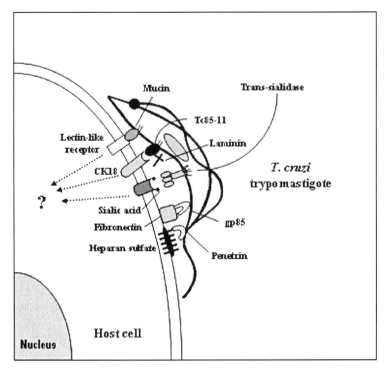

Figure 3. Schematic representation of *T. cruzi*-host cell interactions mediated by parasite molecules with adhesive properties towards target cells and/or extracellular matrix components. Trans-sialidase expressed on the surface of tissue culture trypomastigotes binds to sialic acid of host cell glycoconjugates. It may act as a lectin rather than a sialidase. *T. cruzi* mucin molecules are possibly recognized by lectin-like receptors on the target cell surface. The trypomastigote-specific laminin-binding Tc85-11 putatively binds to target cell cytokeratin 18 (CK18) through a segment downstream of the laminin-binding site. Whether any of these ligand-receptor interactions results in activations of signal transduction pathways is not known. Trypomastigote molecules gp85 and penetrin, with affinity for fibronectin and heparan sulfate/heparin respectively, may contact host cells through these extracellular matrix components.

Acknowledgments

I would like to thank Sergio Schenkman for helpful suggestions and for reading the manuscript.

References

Acosta-Serrano A, Almeida IC, Freitas-Junior LH, Yoshida N, Schenkman S. 2001. The mucin-like glycoprotein super-family of *Trypanosoma cruzi*: structure and biological roles. Mol Biochem Parasitol 114: 143-150.
Burleigh B, Andrews NW. 1995. The mechanisms of *Trypanosoma cruzi* invasion of mammalian cells. Annu Rev Microbiol. 49: 175-200.
Burleigh B, Andrews NW. 1998. Signaling and host cell invasion by *Trypanosoma cruzi*. Curr Opin Microbiol. 1: 461-465.
Caler EV, Vaena de Avalos S, Haynes PA, Andrews NW, Burleigh B. 1998. Oligopeptidase B-dependent signaling mediates host cell invasion by *Trypanosoma cruzi*. EMBO J 17: 4975-4986.
Caler EV, Chakrabarti S, Fowler KT, Rao S, Andrews NW. 2001. The exocytosis -regulatory protein sinaptotagmin VII mediates cell invasion by *Trypanosoma cruzi*. J Exp Med. 193: 1097-1104.

Magdesian MH, Giordano R, Jualiano MA, Juliano L, Schumacher RI, Colli W, Alves MJM. 2001. Infection by *Trypanosoma cruzi*: identification of a parasite ligand and its host-cell receptor. J Biol Chem. 276: 19382-19389.

Ming M, Ewen ME, Pereira MEA. 1995. Trypanosome invasion of mammalian cells requires activation of the TGFβ signaling pathway. Cell 82: 287-296.

Moreno SNJ, Silva J, Vercesi AE, Docampo R. 1994. Cytosolic-free calcium elevation in *Trypanosoma cruzi* is required for cell invasion. J Exp Med 180: 1535-1540.

Pereira MEA, Zhang K, Gong Y, Herrera EM, Ming M. 1996. Invasive phenotype of *Trypanosoma cruzi* restricted to a population expressing trans-sialidase. Infect Immun 64:3884-3892.

Rodriguez A, Martinez I, Chung A, Berlot CH, Andrews NW. 1999. cAMP regulates Ca^{2+}-dependent exocytosis of lysosomes and lysosome-mediated cell invasion by trypanosomes. J Biol Chem 274: 16754-16759.

Ruiz RC, Favoreto S, Dorta ML, Oshiro MEM, Ferreira AT, Manque PM and Yoshida N. 1998. Infectivity of *Trypanosoma cruzi* strains is associated with differential expression of surface glycoproteins with differential Ca^{2+} signaling activity. Biochem J 330: 505-511.

Scharfstein J, Schmitz V, Morandi V, Capella MMA, Lima APCA, Morrot A, Juliano L, Muller-Ester W. 2000. Host cell invasion by *Trypanosoma cruzi* is potentiated by activation of bradykinin B_2 receptors. J Exp Med 192: 1289-1299.

Schenkman S, Eichinger D. 1993. *Trypanosoma cruzi* trans-sialidase and cell invasion. Parasitol Today 9: 218-222.

Schenkman S, Eichinger D, Pereira MEA, Nussenzweig V. 1994. Structural and functional properties of *Trypanosoma* trans-sialidase. Annu Rev Microbiol 48: 499-523.

RECENT DEVELOPMENTS IN THE PATHOLOGY OF CHAGAS DISEASE WITH EMPHASIS ON THE CARDIOVASCULAR SYSTEM

H. B. Tanowitz*, S. M. Factor*, J. Shirani*, A. Ilercil*, M. Wittner*, J. Scharfstein**, L. V. Kirchhoff***
*Albert Einstein College of Medicine, Jacobi Medical Center and Montefiore Medical Center, Bronx, NY
**Universidade Federal do Rio de Janeiro, Brazil
***University of Iowa and Department of Veterans Affairs Medical Center, Iowa City Iowa, Instituto de Biofísica Carlos Chagas Filho

ABSTRACT

Our understanding of the pathology and pathogenesis of Chagas disease has undergone changes over the past several years. Cardiac imaging studies have demonstrated that even in the acute and indeterminate phases of the disease there are significant alterations in cardiac structure and function. These may have both therapeutic and prognostic implications. In recent years Chagas disease has become recognized and an important opportunistic infection in individuals with immunosuppression such as those with HIV/AIDS and as a result of organ transplantation. chagasic heart disease involves both inflammatory and ischemic changes and in recent years the role of the vasculature has received increasing attention. Pathogenic mechanisms have been explored to explain the myocardial dysfunction observed as a result of *T. cruzi* infection. *T. cruzi* infection activates several cardiovascular signaling pathways involving cytokines, nitric oxide, endothelin, kinins and the mitogen activated protein kinases leading to remodeling and cardiovascular dysfunction.

INTRODUCTION

Chagas disease is an important cause of morbidity and mortality in Mexico, Central and South America. In areas where the disease is endemic, it is most common among the rural poor who live in an environment with many feral reservoir hosts. Natural transmission of *Trypanosoma cruzi* to humans is associated with the bite of a triatomine insect, which deposits excreta containing infectious metacyclic trypomastigotes that then contaminate the bite site or mucosal surfaces. Blood transfusion also has become an important mode of transmission, and transfusion Chagas disease has been reported in non-endemic areas. Chronic symptomatic cases of Chagas disease have become evident in non-endemic areas as thousands of immigrants from Mexico, South and Central America have moved to the United States, Canada and Europe. Five autochthonous human cases have been reported in the United States. Congenital transmission, breast-feeding, organ transplantation and laboratory accidents are other modes of transmission (Tanowitz et al., 1992a).

Host cell entry by the parasite is a complex process involving many host and parasite factors. Trypomastigotes released from infected host cells may infect adjacent uninfected cells and/or enter the bloodstream and

Tanowitz et al.

lymphatics infecting distant tissues. No tissue is spared from infection but *T. cruzi* strains may vary in tissue tropism. The cells of the mononuclear phagocytic, nervous, and muscular (striated and cardiac) systems appear to be particularly vulnerable.

Figure 1. A. Autopsy specimen from a patient with chronic chagasic cardiomyopathy. There is

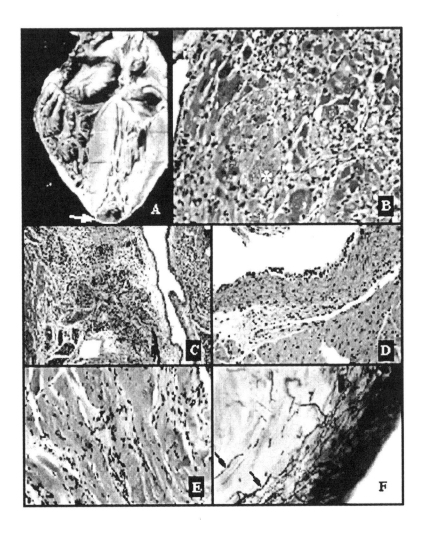

four-chamber enlargement of the heart due to dilatation and hypertrophy. There is also an apical aneurysm (arrow). *Reprinted with permission of the Armed forces Institutes of Pathology, Washington, D.C. B.* Acute myocarditis in a CD1 mouse infected with the Brazil strain of *T. cruzi.* There is inflammation and myonecrosis. Parasite pseudocysts are evident (*). C. Ganglionitis in the myocardium of a mouse acutely infected with *T. cruzi.* D. Endotheliititis of the aorta from a mouse with acute *T. cruzi* infection. E. Chronic cardiomyopathy in a mouse infected for 150 days with the Brazil strain of *T. cruzi.* Note the inflammation and fibrosis. F. Microvasculature spasm and aneurysm formation (arrows) in the myocardium of a CD1 mouse infected with the Brazil strain of *T. cruzi.*

CLINICAL CHAGASIC HEART DISEASE

In natural transmission, at the site of parasite entry, an inflammatory lesion (chagoma) may develop. The inflammatory process spreads regionally and local lymphadenopathy may be evident. Asynchronous cycles of parasite multiplication, cell destruction and reinfection occur within cells of the mononuclear phagocytic system. Most individuals with acute Chagas disease have only mild symptoms. However, children and, less frequently, adults may develop severe symptoms after an incubation period of 7 to 14 days. These symptoms include unilateral, painless, periorbital edema (Romaña's sign). Other manifestations include fever, lymphadenopathy, hepatosplenomegaly, nausea, vomiting, diarrhea, rash, anorexia, lassitude, and meningeal irritation. A small number of patients with acute Chagas disease develop severe myocarditis. Myocarditis may be clinically evident and include tachycardia, cardiomegaly and congestive heart failure (CHF). The ECG may reveal prolongation of the P-R interval, nonspecific T-wave changes, and low voltage. Abnormalities such as right bundle branch block (RBBB), left anterior fascicular block, atrioventricular block or low QRS voltage have been noted in 40% of patients studied during the acute febrile illness. The appearance of arrhythmias, heart block, or progressive CHF during the early phase of acute Chagas disease is an indicator of a poor prognosis.

Trypomastigotes may be found in the peripheral blood and cerebrospinal fluid (CSF) during acute infection. Anemia, thrombocytopenia, leukocytosis with a lymphocyte prominence, elevated liver and cardiac enzymes are among common laboratory abnormalities reported. A small percentage of individuals with acute disease die of complications associated with acute myocarditis or meningoencephalitis. Nearly all patients with acute infection recover completely within 3 to 4 months, and later have no recollection of an acute illness. In these individuals the parasitemia has become so low that evidence of infection must rely on other diagnostic approaches (i.e., serology, PCR, hemoculture). In those who die during the acute phase, amastigotes are easily demonstrated in many tissues. Those who survive the acute illness enter what is referred to as the indeterminate phase in which they are asymptomatic, but careful clinical and pathologic studies demonstrate that the disease may be progressing.

The clinical manifestation of chronic chagasic heart disease (CHD) may occur years or decades after acute infection. A recent report from a hospital in Brazil identified Chagas heart disease in 11% of patients who died from a cardiovascular cause (de Oliveira et al., 2000). Chronic CHD presents insidiously or abruptly with arrhythmias, dilated congestive cardiomyopathy, thromboembolic phenomena and sudden death. Apical aneurysm, with or without thrombus formation is one of the hallmarks of this disease and is frequently, though not exclusively, found in the left ventricle (LV) (Oliveira et al., 1981). Although there is usually an absence of significant major coronary artery lesions, several case reports of myocardial infarction in individuals with Chagas disease have been reported. Thromboembolization to the coronary arteries is thought to occur but coronary thrombosis is rare. Destruction of conduction tissue results in the conduction abnormalities.

In endemic areas the presence of RBBB, associated with an anterior fascicular block is highly suggestive of CHD. Conduction defects are not uncommon and may necessitate pacemaker placement. Asymptomatic seropositive individuals have a high prevalence of ECG abnormalities. There is increased mortality in seropositive individuals with RBBB and anterior fascicular block or ventricular extrasystoles. It has been difficult to determine which electrophysiological abnormalities predict progression to severe cardiomyopathy. An autopsy study of more than 5,000 patients with Chagas disease demonstrated no relation between age and duration of disease or the presence of an apical aneurysm (Oliveira et al., 1981). In addition, it was found that 30 % of asymptomatic patients with normal ECG and radiological findings had normal myocardial biopsies. The incidence of abnormal biopsies increased in asymptomatic patients with abnormal ECG or radiographic findings and in clinically symptomatic patients. Of those histopathological parameters examined, myocardial hypertrophy and fibrosis but not inflammation correlated best with the severity of cardiac symptoms (Pereira-Barretto et al., 1986).

CARDIAC IMAGING: CLUES TO PATHOLOGY

Echocardiography has evolved as an important modality in evaluating the cardiac status of patients with CHD and the resulting data have provided important information regarding clinical-pathological correlation. During acute infection echocardiography may reveal abnormal segmental LV wall motion, most commonly in the anterior and/or apical regions (Parada et al., 1997). The overall LV size and ejection fraction is typically normal. Fatal acute myocarditis may occur in a minority (5-10%) of patients. The acute phase may be followed by a long indeterminate phase during which ECG or radiological abnormalities may not be detectable. A substantial number of these individuals (10-30%) will eventually evolve into the serious and potentially fatal chronic phase of the disease.

Early echocardiographic studies focused on evaluating patients with symptomatic advanced CHD (Acquatella et al., 1980). The most typical finding in such individuals is an apical LV aneurysm with or without thrombus, and/or basal inferoposterior hypokinesis or akinesis with preserved contractile function in the ventricular septum (Acquatella et al., 1980). The left atrium is often enlarged and the LV is dilated with reduced systolic function. In cases of advanced cardiomyopathy with CHF, biventricular dilatation occurs without increased wall thickness (eccentric hypertrophy). The segmental and global LV dysfunction occurs in the absence of epicardial coronary artery disease. Although a number of studies have demonstrated an overall good prognosis for patients in the indeterminate phase of CHD, in at least one area 2 to 5% of patients progress annually into the chronic phase and some patients may die suddenly (Acquatella et al., 1980). Therefore identification of occult cardiac dysfunction may be of clinical significance for risk stratification. Cardiac abnormalities have been demonstrated in autopsy specimens or by endomyocardial biopsy in those with no apparent cardiomyopathy on noninvasive testing. Pathological findings have included carditis, focal inflammatory infiltration, myocytolysis and altered interstitial collagen matrix.

Cardiac structural and functional alterations are detectable in some patients in the indeterminate phase of CHD, when sensitive methods of assessment are used such as treadmill exercise testing, high resolution and dynamic ECG, noninvasive autonomic nervous system evaluation, radioisotope ventriculography and echocardiography. The goal of such studies is to identify prospectively *T. cruzi*-infected persons who are at an increased risk of sudden death or progression to the chronic symptomatic phase. Abnormalities of LV contraction and relaxation have been demonstrated on simultaneous phonocardiography, apexcardiography and echocardiography in asymptomatic patients with positive serology, without evidence of cardiac disease. Employing pulsed-wave and tissue Doppler imaging techniques, a recent study identified an abnormally prolonged isovolumic relaxation time and deceleration time of the early transmitral flow (Barros et al., 2001). In addition, the isovolumic contraction time was prolonged in the ventricular septal myocardium. These observations indicate that subtle abnormalities of both diastolic and systolic LV function are present in the indeterminate phase.

Further support for the presence of occult LV dysfunction has been presented by a study using dobutamine stress echocardiography to evaluate contractile reserve in patients with and without resting wall motion abnormality (Acquatella et al., 1999). Neither group of patients reached target heart rate during dobutamine infusion, a finding indicative of chronotropic incompetence. In addition, both groups of patients exhibited a blunted contractile response to inotropic stimulation. Thus, compared with a normal value of 67%, the fractional LV area change was only 50% in those with normal and 30% in those with abnormal resting wall motion. A biphasic segmental wall motion, defined as an initial improvement followed by deterioration at higher doses of dobutamine, was only seen in individuals with abnormal LV wall motion at baseline (Acquatella et al., 1999). The latter was most commonly seen in the mid-posterior or posteroinferior regions of LV. All patients had normal epicardial coronary arteries by angiography. This may reflect abnormalities in coronary flow reserve due to microcirculatory dysfunction. In fact, impaired endothelial-dependent coronary vasodilation has been demonstrated in those with normal coronary arteries on angiography. Myocardial perfusion defects have been shown by thallium-201 scintigraphy in individuals with and without apparent cardiac involvement. These perfusion abnormalities closely parallel those of [I-123]-meta-iodobenzoguanidine (MIBG) scintigraphic findings used for evaluation of cardiac sympathetic innervation (Simoes et al., 2000). Regional myocardial ischemia, or inflammation, or both, may be responsible for these scintigraphic abnormalities.

Prognostic information has been provided by echocardiography in patients with established CHD. In this regard, both LV size and systolic function have been found to predict short-term prognosis. The prognosis of cardiomyopathy has also been found to be worse than that for other etiologies of LV systolic dysfunction. This is attributed to the presence of an apical aneurysm that alters the normal prolate-ellipsoid shape of the LV and has a potential for producing arrhythmias, thromboembolic events, or CHF (Bestetti and Muccillo, 1997). More recently, cardiac MRI has been used in the evaluation of patients with CHD and may prove to be a useful adjunct. Myocardial inflammation can also be detected by means of gallium-67 scintigraphy and myocardial tissue characterization using MRI with contrast. Finally, both MRI and echocardiography have been employed in the murine model of CHD where serial studies post-infection have yielded important results regarding structural and functional abnormalities (Huang et al., 1999b).

PATHOLOGY OF CHAGASIC HEART DISEASE (CHD)

CHD represents the interplay of inflammation and ischemia as well as ischemia/reperfusion. During acute infection foci of myonecrosis, myocytolysis and vasculitis, accompanied by an inflammatory exudate consisting primarily of leukocytes, are observed. However, we have observed eosinophils and mast cells in the mouse model. Pseudocysts containing amastigotes can be found interspersed among the degenerating fibers.

T. cruzi gains access to the cardiac myocytes by first invading endothelial cells, the interstitial areas of the vascular wall, and the myocardium. Parasites can be seen in and around the endothelium of infected mice. Trypomastigotes pass two basal laminae areas and two layers of extracellular matrix (ECM). Parasite enzymes such as proteases, gelatinases and collagenases degrade native type I collagen, heat denatured type I collagen and native type IV collagen. Proteolytic activities against laminin and fibronectin have also been detected. This has also been observed with cruzipain, which also cleaves collagen IV and heparan sulfate proteoglycans (JS, unpublished observations). These enzymes may play an important role in degradation of ECM and subsequent parasite invasion. It has been proposed that degradation of the collagen matrix, evident in acute murine Chagas disease, may result in chronic pathology such as apical thinning of the LV.

There are three layers of cardiac myocytes in the heart that are obliquely oriented to each other and meet at the apex. When there is damage, such as that caused by ischemia and necrosis, matrix degradation occurs leading to a slippage of the ventricular layers with mural thinning and aneurysm formation. Damage to this area of the heart and remodeling of the wall is frequently encountered in CHD. Remodeling, in the context of the cardiovascular system, refers to the structural changes associated with inflammation, necrosis, hypertrophy and ventricular dilation. In the course of chronic CHD myonecrosis, myocytolysis and contraction band necrosis are evident. The later results from transient hypoperfusion followed by reperfusion, such as after local spasm of the coronary microvasculature. Focal and diffuse areas of myocellular hypertrophy may be observed with or without inflammatory infiltrates. In other areas, focal fibrosis replacing previously damaged myocardial tissue is evident. An important feature of CHD is a dense accumulation extracellular collagen that encloses fibers or groups of fibers. All areas of the heart, including the conduction pathways, may be involved. Microvascular involvement, manifested by basement membrane thickening, has been demonstrated. These irreversible changes lead to functional disturbances. The remodeling process results in damage to the ECM and the replacement of cardiac myocytes and cells of the vasculature by fibrous tissue (Higuchi et al., 1999). This results in thinning of the myocardium and hypertrophy of the remaining cardiac myocytes.

Cellular proliferation in the myocardium involving inflammatory cells, fibroblasts, endothelial cells and smooth muscle cells is observed and contributes to the pathogenesis of CHD. It should be noted that both hypertrophy and proliferation characterize CHD. Chronic CHD is characterized pathologically by focal T lymphocyte infiltrates (predominantly CD8+), myocyte necrosis and marked reactive and reparative fibrosis. Recently, apoptosis has been evaluated in the myocardium in CHD. In humans apoptosis was not evident in cardiac myocytes (Rossi and Souza, 1999). However, in infected dogs, it was observed in cardiac myocytes, endothelial cells and inflammatory infiltrates (Zhang et al., 1999).

Observations from our laboratory in acutely infected mice are in agreement with the findings in dog, except that we did not observe evidence of apoptosis in cardiac myocytes. Therefore, both apoptosis and necrosis may coexist in chagasic hearts. However, more experimentation is necessary. The apparent absence of parasites in the myocardium during routine histologic examination in chronic CHD, even in the presence of marked morphological and functional changes, has engendered much speculation as to the possible pathogenic etiologies. Parasite persistence, autoimmunity and disturbances in the autonomic nervous system (ANS) have emerged as possible important factors. These will be discussed elsewhere in this volume.

The autonomic dysfunction observed in *T. cruzi* infection has received limited attention as a possible cause of cardiac and gastrointestinal manifestations of chronic disease. The reported depletion of cardiac acetylcholine and choline acetyltransferase observed in experimental animals could adversely affect parasympathetic nervous function. Dysfunction of the ANS led to the idea that chronic cardiomyopathy may be caused by autonomic neuronal damage in the heart. Studies demonstrating destruction of cardiac ganglia also support this view. However, extreme variability in the density of myocardial vagal ganglia of chagasic hearts has been reported, in part reflecting the limits of pathological and histochemical analysis. Since vagal denervation has been observed in dilated cardiomyopathies of other etiologies, its association with advanced Chagas disease may not imply a primary pathogenic role. Of note is the fact that alterations in parasympathetic function can be demonstrated in seropositive chagasic patients before the onset of clinically significant myocardial dysfunction. This may be a unique aspect of the disease. Abnormalities in the sympathetic component of the ANS in chagasic patients, as reflected in the reduction in levels of plasma norepinephrine were also reported, in contrast to elevated levels in non-chagasic cardiomyopathic patients. This is an area of investigation that is ongoing in several laboratories.

THE VASCULOPATHY OF CHAGAS DISEASE

An overlooked aspect of the pathology of CHD has been the consequence of infection on the vasculature. For example, amastigotes are observed in endothelial cells lining the coronary microcirculation in association with endothelial damage. In animal models and in humans there is an intense vasculitis of the aorta, coronary arteries, smaller myocardial vessels and the endocardial endothelium. In addition, a vasculitis is observed in other organs, such as the liver, indicating that the *T. cruzi*-induced vasculopathy is not limited to the heart and that a vasculopathy is a part of the systemic pathogenesis of *T. cruzi* infection.

Murine *T. cruzi* infection was found to be associated with segmental microvascular spasm and microaneurysm formation as well as reduced red blood cell velocity in vascular beds. This was not observed in infected mice treated with verapamil, or in uninfected mice. Coronary vascular perfusion was found to be reduced in infected mice. Observations in experimental animals demonstrated platelet thrombi, increased platelet aggregation and elevated plasma levels of thromboxane A_2, which promotes vasospasm and platelet aggregation. Further evidence of the involvement of the microvasculature in the pathogenesis of CHD was demonstrated by our studies with verapamil. Administration of this drug to *T. cruzi*-infected CD1 mice decreased mortality and reduced myocardial inflammation and fibrosis.

Our data and those of others indicate that verapamil acts on vascular smooth muscle cells (Petkova et al., 2001).

Infected endothelial cells exhibit perturbations in the host-cell signal transduction pathways that are responsible for endothelial dysfunction. *In vitro* and *in vivo* studies demonstrated that *T. cruzi*-infection of the endothelium resulted in the activation of the NF-κB pathway and the increased synthesis of cytokines and vascular adhesion molecules, important components of the inflammatory response (Huang et al., 1999a). *T. cruzi*-infection of endothelial cells caused increased synthesis of the vasoactive peptide, endothelin-1 (see below). These and other infection-associated perturbations in endothelial-cell signal transduction mechanisms may contribute to focal pathology, which may include the coronary microvascular spasm reported in acute *T. cruzi* infection in mice.

PATHOLOGY IN IMMUNOCOMPROMISED PATIENTS

In immunocompromised persons, chronic *T. cruzi* infection can become reactivated and variably lead to signs and symptoms characteristic of acute Chagas disease. This phenomenon has been noted in patients with chronic *T. cruzi* infections who are given immunosuppressive drugs after organ transplantation. Although a few patients with renal transplants have developed reactivated *T. cruzi* infection, the most noteworthy individuals in this group are those who have undergone cardiac transplantation (Bocchi et al., 1993). Recrudescent Chagas disease has also been reported in several dozen *T. cruzi*-infected patients who were immunosuppressed by HIV (Corti, 2000; Rocha et al., 1993; Sartori et al., 1998; Sartori et al., 1999).

Recrudescence of chronic *T. cruzi* infection in HIV-infected patients

When patients chronically infected with *T. cruzi* are immunosuppressed by HIV, the parasitosis can reactivate and cause serious and even life-threatening disease. Although several dozen such cases have been reported to date, it should be noted that the incidence of clinically significant reactivation of *T. cruzi* in patients with HIV is low, as epidemiologic data indicate clearly that there are many thousands of such dually-infected people. It is reasonable to assume that many patients in the latter group who do not have clinical findings attributable to *T. cruzi* infection do have parasitemia higher than those typically found in parasitized immunocompetent patients. This issue has not been studied, however, and thus little is known about the early phase of *T. cruzi* reactivation.

The clinical picture in HIV patients with reactivated *T. cruzi* infection can include fever, malaise, anorexia, and myalgias. In addition, the majority of the patients with dual infections reported to date have had signs and symptoms resulting from lesions in the central nervous system and/or the heart, which occur with approximately equal frequency. Neurologic findings reflect increased intracranial pressure, meningitis, encephalitis, and inflammation involving the cranial nerves, and have included mental status changes, cranial nerve dysfunction, paresthesias, headache, focal motor deficits, and seizures. In patients with neurologic findings due to cerebral *T. cruzi* lesions, the CSF usually shows a lymphocytic pleocytosis and an elevated protein level. Moreover, in these patients parasites usually can be found in both the CSF and the blood. Imaging studies of the cerebrum usually show single or multiple ring-enhancing pseudotumors surrounded by edema, typically located in the cortex or subcortical white matter. This localization would be atypical for lesions of *Toxoplasma* encephalitis, which are usually found in the thalamus and the basal ganglia. However, imaging studies are not definitive.

Gross examination of the brains of HIV-infected patients with cerebral *T. cruzi* lesions shows increased weight and blunting of the sulci. The picture is that of focal meningoencephalitis or encephalitis. The pseudotumors have a soft consistency and they can be be several cm. in diameter. As noted, the cortex and subcortical white matter are most heavily affected, but smaller lesions also occur in the brain stem and cerebellum. It is important to note that such lesions do not occur in immunocompetent persons with chronic *T. cruzi* infections. Focal hemorrhage and necrosis are variably present. Microscopic examination shows widespread parenchymal and perivascular necrosis as well as exudate containing mononuclear cells and granulocytes. Intracellular amastigotes are found in macrophages, glial cells, and epithelial cells, and extracellular trypomastigotes are often also present.

Reactivation of *T. cruzi* infection in carriers of HIV can also affect the heart and cause clinically apparent acute myocarditis (Sartori et al., 1998). In the small number of such patients described to date, findings have included CHF and a variety of arrhythmias. Globally increased heart size was noted in three patients examined. Microscopically, diffuse and focal mononuclear cell infiltration of myocardial tissue has been noted. Focal degeneration and necrosis of cardiac myocytes and interstitial edema also have been observed. Numerous pseudocysts consisting of amastigote-filled cardiomyocytes are present in a focal distribution.

Post cardiac transplantation reactivation of chronic *T. cruzi* infection

Soon after the first heart transplants were performed in *T. cruzi*-infected patients in the 1980s the risk of recrudescence of the parasitosis became apparent. This occurrence was not surprising, as reactivation of chronic *T. cruzi* infections had been observed previously in other iatrogenically immunosuppressed persons. The frequency of reactivation observed after heart transplantation is variable, depending on the intensity of the immunosuppression and the diligence with which parasites are sought. The most common manifestations of *T. cruzi* reactivation in this patient population are fever, myocarditis in the new heart, and subcutaneous nodules. From a parasitologic perspective, circulating parasites are generally easier to detect in this group than in immunocompetent patients, whatever the detection method is used. Histologic examination of tissue obtained by endomyocardial biopsy after transplantation shows mononuclear infiltration, interstitial edema, and scattered pseudocysts, similar to the findings observed in cardiac tissue from HIV patients with reactivated *T. cruzi* infection. The subcutaneous nodules occasionally observed in these patients can ulcerate and have been mistaken for lesions of cutaneous leishmaniasis. On histologic examination large numbers of intracellular amastigotes can be seen which are morphologically indistinguishable from amastigotes of *Leishmania* species. Parasite-containing cutaneous lesions have never been reported in immunocompetent persons who chronically harbor *T. cruzi*. An association between the recrudescence of *T. cruzi* infection and episodes of rejection of the transplanted hearts has not been established.

Tanowitz et al.

NEW INSIGHTS REGARDING THE MOLECULAR PATHOGENESIS OF CHAGASIC HEART DISEASE

The mouse model of CHD has been utilized because it has been generally acknowledged that despite its pitfalls it faithfully recapitulates many of the aspects of human disease. Much of the focus has been on the acute infection because it is thought by many that early events set the stage for chronic pathology and dysfunction. However, many investigators have disputed this contention. In recent years, alterations in cardiovascular signaling pathways have been found to contribute to the pathogenesis of CHD and the remodeling of the myocardium resulting in chronic cardiomyopathy. In addition, the role of the microvasculature in CHD has now been established (Petkova et al., 2001)

INFLAMMATORY MEDIATORS

The inflammatory process is a critical factor in the pathogenesis of CHD. Several studies in experimental animals and in humans, as well as *in vitro* studies, have demonstrated that *T. cruzi* infection is associated with increased expression/activation of a number of important inflammatory mediators. For example, infection of endothelial cells resulted in activation of NF-κB, increased synthesis of cytokines and induction of vascular adhesion molecules (Tanowitz et al., 1992b; Huang et al., 1999a). Increased expression of adhesion molecules has also been reported in the myocardium from infected animals and humans (Reis et al., 1993). The expression of myocardial cytokines and chemokines is significantly increased in infected rodent models (Aliberti et al., 2000). In addition, an increased expression of inducible and endothelial nitric oxide synthase (iNOS and eNOS) has been reported in experimental acute CHD (Huang et al., 1999b).

The importance of the cytokine-iNOS pathway in *T. cruzi* infection has been investigated. Cytokines induce the iNOS gene and are accompanied by the generation of NO, which has anti-parasitic activity. To assess more directly the role of NO in the pathogenesis of CHD, *T. cruzi* infection in iNOS knock out (KO) mice was investigated. Neither the iNOS KO or syngeneic wild type mice died acutely when infected with the Brazil strain. Both infected iNOS KO and wild type mice developed cardiac pathology, including right ventricular dilatation and decreased fractional shortening. The degree of change, however, was markedly less in the iNOS KO mice. These data are consistent with the notion that NO generated from iNOS contributes to the structural and functional alterations observed in acute murine Chagas disease. Importantly, cytokines and NO have been associated with the cardiac dysfunction associated with myocarditis and cardiomyopathies of diverse etiologies. The action of inflammatory mediators, including NO, has traditionally been viewed as a double-edged sword, i.e.; it contains the infection but also harms the host. Chagas disease is an example of this because NO is needed for parasite killing but increased NO leads to myocardial structural and functional abnormalities. In addition, recent evidence suggests that NO may also be critical in the pathogenesis of the autonomic dysfunction associated with *T. cruzi* infection.

Endothelin-1 (ET-1)

ET-1, a 21-amino acid peptide, is a potent vasoconstrictor. Cardiac myocytes and endothelial cells are among the cells that synthesize ET-1. Locally produced ET-1 acts on cardiac myocytes in both an autocrine and/or paracrine manner, increasing the contractility of smooth muscle cells and chronically inducing myocardial hypertrophy and cardiac myocyte injury. ET-1 production was increased in the myocardium of rats with CHF. In the setting of myocardial infarction, increased plasma ET-1 levels correlated with infarct size and levels of ET-1 correlated with the severity of CHF. ET-1 levels reflect the degree of endothelial and myocardial cell damage as well as a possible mechanism of myocardial damage. Importantly, treatment with an ET_A receptor antagonist improved the survival of animals with CHF and was accompanied by improvement in both LV dysfunction and remodeling, suggesting that upregulation of ET-1 may be a potential target for therapeutic intervention in the treatment of CHF.

Studies on the putative role for ET-1 in the modulation of cardiovascular structure have centered on the role of ET-1 in the induction of smooth muscle cell (SMC) proliferation. For example, following balloon angioplasty (a well-studied model of vascular damage), there is neointima formation accompanied by increased tissue levels of ET-1. This was reported to be associated with induction of mRNAs for ET-1, endothelin converting enzyme (ECE), and the ET_A receptor. The stimulation of SMC hypertrophy and proliferation by ET-1 is mediated at least in part, by activation of the SMC ET_A receptor. Activation of this receptor results in the synthesis of Type I and III collagens and reduction in collagenase activity. These observations could account for some of the vascular remodeling observed as a result of cardiovascular damage occurring during acute Chagas disease. Although ET-1 was previously regarded solely as a vasoactive agent, it is now recognized that it also can act as a pro-inflammatory cytokine. For example, ET-1 stimulates monocytes to produce cytokines and induces the expression of vascular adhesion molecules. High levels of ET-1 are found in alveolar macrophages, leukocytes and fibroblasts. Collectively, these observations indicate that ET-1 is closely associated with the inflammatory process.

During the course of acute murine *T. cruzi* infection the plasma levels of ET-1 have been shown to be elevated and this was accompanied by increased myocardial expression of mRNAs for preproET-1 (precursor), ECE, and ET-1. Immunohistochemistry (IHC) studies revealed increased expression of ET-1 protein in the endothelium of conduit vessels (i.e., aorta and coronary arteries), capillaries and the endocardial endothelium. Many of the parasitized and necrotic cardiac myocytes also stained strongly with anti-ET-1 antibodies. Cardiac myocytes, fibroblasts and inflammatory cells synthesize ET-1. Therefore, the increase in ET-1 observed in the myocardium of *T. cruzi*-infected mice is likely to have several sources. The increase in ET-1 is a reflection of endothelial cell and cardiac myocyte damage as well as a mechanism to explain, in part, the vascular spasm and ischemia observed as a consequence of this infection. *T. cruzi* infection of cultured endothelial cells resulted in the increased synthesis of biologically active ET-1.

In order to examine directly the role of ET-1 in an animal model of *T. cruzi* infection, we infected mice in which the ET-1 gene was deleted from cardiac myocytes [(ET-1 (flox/flox); α-MHC-Cre (+)] with the Brazil strain of the parasite. In these mice the infection-associated increase in right ventricular enlargement and decrease in percent fractional shortening was significantly attenuated in comparison to control mice, as determined by cardiac MRI and echocardiography. In additional studies, we demonstrated that treatment of infected mice with phosphoramidon, an inhibitor of ECE, resulted in improved myocardial function as compared to infected untreated mice. Collectively, these data are consistent with the concept that ET-1 contributes to the pathogenesis of CHD.

ACTIVATION OF THE KININ SYSTEM BY *T. CRUZI* KININOGENASES (CRUZIPAIN)

Cruzipain is a developmentally regulated cysteine proteinase that has recently been validated as a chemotherapeutic target of Chagas disease (Engel et al., 1998). Importantly, in an autopsy IHC study of the myocardium of individuals with chronic CHD, monoclonal antibodies to cruzipain revealed presence of antigen deposits in areas infiltrated by inflammatory cell (Morrot et al., 1997). Since cruzipain is expressed at high levels by amastigotes, these deposits most likely resulted from amastigote remnants, rather than intracellular pseudocysts, which are rarely detected in tissues from chronically infected patients. Previous analysis of the immune response profile of chagasic patients indicated that cruzipain induced potent humoral and T cell responses that may contribute to the pathology. In addition, recent studies indicated that trypomastigotes rely on cruzipain to activate the pro-inflammatory kinin cascade system (Scharfstein et al., 2000; Lima et al., 2002)

The term "kinin" refers to a small group of vasoactive metabolites structurally related to the nonapeptide bradykinin, the generation of which is dependent on the proteolytic excision of the internal kinin segment from the plasma precursor proteins, H and L-kininogens. As a result of vascular injury, activated forms of plasma kallikrein release the nonapeptide bradykinin from H-kininogen. During inflammation, tissue kallikrein liberates lysyl-bradykinin (kallidin) from blood-borne L-kininogen. In other settings, Met-kallidin is liberated from oxidized kininogens due to the concerted action of neutrophil elastase and mast cell tryptase. The biological function of the short-lived kinins is mediated via G-protein coupled receptor subtypes expressed at high levels by vascular endothelial and smooth muscle cells. The primary kinin products, bradykinin or lysyl-bradykinin stimulate B2, a receptor subtype constitutively expressed by cardiovascular cells, while B1, a receptor upregulated by various cell types during inflammation, is triggered by their metabolites, des-Arg kinins. The half-life and pharmacological specificity of the kinin agonists is tightly regulated by metallopeptidases. For example, the agonist activity of kinin peptides is destroyed by kininase II (ie., the angiotensin-converting enzyme, ACE). Kinase I, on the other hand, removes the C-terminal Arg from the primary kinin product, generating the B1-agonist (des-Arg kinins).

Analysis of the substrate specificity of cruzipain revealed striking similarities with tissue kallikrein. It was further demonstrated that trypomastigotes mobilize cruzipain to activate endothelial cells through the B2-kinin-receptor subtype (Scharfstein et al., 2000). The activation of this receptor occurs upon parasite attachment to target cells that display high levels H-kininogen at their cell surfaces. The efficiency of kinin release reaction is drastically enhanced by mutual cooperative interactions between heparan sulfate proteoglycans, H-kininogens and cruzipain (Lima et al., 2002). The signaling activity of the released kinins is then modulated by kinin-degrading peptidases (eg. ACE). Importantly, the vigorous $[Ca^{2+}]_i$ transients relayed through the B2-kinin receptor markedly increased the susceptibility of endothelial cells to parasite invasion. Analysis of the host cell preference of trypomastigotes that overexpressed different cruzipain isoforms indicated that the kinin-activating phenotype was linked to expression of particular subsets of isoenzymes. Collectively, these studies suggest that differences in the repertoire of cruzipain expressed by the heterogeneous *T. cruzi* clones may influence parasite virulence and/or pathogenicity.

Although traditionally viewed as factors that participate in acute inflammation (eg. edema, vasodilation, and nociception), the kinin system is presently also regarded as a key modulator of circulatory homeostasis. For example, recent studies suggest that vasoactive kinins exert cardioprotective effects in experimental models of ischemia/reperfusion. Hence, it is possible that activation of the kinin system by extracellular parasites may to be of mutual advantage to the parasite-host relationship, since it may accelerate repair mechanisms in the injured myocardium, while at the same time maximizing opportunities for parasite infection and proliferation in the heart.

Recently, the interaction of trypomastigote with the vascular endothelium was examined by intravital microscopy. Using the hamster cheek pouch (HCP) as a model, permeability increases mediated by kinin-receptors were observed upon topical application of living trypomastigotes (Todorov et al., unpublished data). Since normal HCP tissues (i.e., non-inflamed) do not accumulate sufficient quantities of kininogens, the experiments performed with living trypomastigotes suggest that the influx of the kinin-precursor proteins from the bloodstream into extravascular tissues is a pre-requisite for efficient activation of the kinin cascade. Recent analysis of the vascular responses which trypomastigotes elicit in wild type mice, or in KO mice with targeted deletion of kinin-receptors strengthen this concept.

Since kinin-receptors and ACE play key roles in cardiovascular homeostasis, it is possible that parasite-induced pathology may be critically influenced by fluctuations in the levels of kinin-degrading peptidase, which may vary from one individual to another. Interestingly, it was observed that the kidney parenchyma and lungs, i.e., highly vascularized tissues that abundantly express ACE, are usually free from parasites. However, massive infection and inflammation develops in other organs. These findings suggest the possibility that that high-level expression of ACE or other kinin-degrading peptidases may protect the microvasculature from infection and from other pathological responses mediated by B2 kinin-receptors, such as edema, NO production, and PGE2 release. Moreover, it is possible that the delicate host-parasite balance achieved during the indeterminate stage of infection may be disturbed as result of a excessive kinin-release resulting from a marked decrease in myocardial levels of ACE (or other kinin-degrading peptidases). Alternatively, increased expression of kininase I may generate high levels of the agonist (des-Arg bradykinin or des-Arg kallidin) for B1, the receptor subtype that is upregulated during inflammation.

Recent investigations indicate that signaling through the B1 receptor represents another potential pathway for parasite invasion and proliferation in cardiovascular cells. Maintenance of the host-parasite balance may thus depend on robust activation of cellular immunity, a process that is inevitably linked to excessive production of inflammatory cytokines. Furthermore, triggering of the endothelial cell B1 receptor by *T. cruzi* clones that activate the kinin cascade may lead to the transcriptional activation of NF-κB, a pathway that upregulates the expression of vascular adhesion molecules. It is likely that continued studies of the mechanisms underlying activation of the kinin cascade by the parasites will yield useful information in understanding the molecular pathogenesis of CHD.

MITOGEN-ACTIVATED PROTEIN KINASES (MAPKS) AND CYCLINS

There are three major MAPKs in mammalian cells, extracellular signal regulated kinase (ERK), c- JunNH$_2$-terminal kinase (JNK) and p38 MAPK. These contribute to the induction of transcription factors AP-1 and NF-κB. Important downstream targets of the MAPKs and ET-1 pathways are the cyclins, in particular cyclin D1. Injury to the cardiovascular system caused by hypoxia, infection, ischemia/reperfusion and restenosis after balloon angioplasty activates these important signaling pathways that have been implicated in the pathogenesis of cardiovascular disease.

In the myocardium of infected mice we found an activation of ERK and AP-1 (c-jun and c-fos). IHC revealed expression of c-jun and c-fos protein in the myocardium especially in the vasculature (Huang et al., 2000). In addition, there was an induction of NF-κB. Induction/activation of ERK and NF-κB causes increased expression of cyclin D1. All of these factors are responsible for cell proliferation. Indeed, western blot demonstrated increased expression of myocardial cyclin D1 and proliferating cellular nuclear antigen (PCNA) in infected mice. IHC studies revealed that staining for cyclin D1 and PCNA was most intense in the vascular and endocardial endothelium, inflammatory cells and the interstitium (fibroblasts). The induction of these proteins is important to the remodeling process in the cardiovascular system. These observations may provide targets of potential adjunctive therapy to ameliorate cardiovascular disease.

T. cruzi infection causes ischemic and inflammatory changes resulting in cardiomyopathy and vasculopathy. There are perturbations in several important signaling pathways that include cytokines, chemokines, ET-1, kinins, MAPKs, the transcription factors AP-1 and NF-κB as well as cell cycle regulatory proteins. These events lead to abnormalities in vascular function as well as cellular proliferation in the cardiovascular system. The perturbations result in remodeling and cardiomyopathy. Delineation of these pathways may provide further understanding of the nature of the host-parasite relationship and will provide insights for adjunctive therapy for the treatment of Chagas disease.

References

Acquatella H, Perez JE, Condado JA, Sanchez I. 1999. Limited myocardial contractile reserve and chronotropic incompetence in patients with chronic Chagas disease. J Am Coll Cardiol 33:522-529.

Acquatella H, Schiller NB, Puigbo JJ, Giordano H, Suarez JA, Casal H, Arreaza N, Valecillos R, Hirschhaut E. 1980. M-mode and two-dimensional echocardiography in chronic Chagas heart disease: a clinical and pathologic study. Circulation 62:787-799.

Aliberti JC, Souto JT, Marino AP, Lannes-Vieira J, Teixeira MM, Farber J, Gazzinelli RT, Silva JS. 2001. Modulation of chemokine production and inflammatory responses in interferon-γ-and tumor necrosis factor-R1-deficient mice during *Trypanosoma cruzi* infection. Am J Pathol 158:1433-1440.

Barros MVL, Rocha MOC, Ribeiro ALP, Machado FS. 2001. Doppler tissue imaging to evaluate early myocardium damage in patients with undetermined form of Chagas disease and normal echocardiogram. Echocardiography. 18:131-136.

Bestetti RB, Muccillo G. 1997. Clinical course of Chagas heart disease: a comparison with dilated cardiomyopathy. Int J Cardiol 60:187-193.

Bocchi EA, Bellotti G, Uip D, Kalil J, Higuchi M, Fiorelli A, Stolf N, Jatene A, Pilleggi F. 1993. Long-term follow-up after heart transplantation in Chagas disease. Transplant Proc 25:329-330.

Corti M. 2000. AIDS and Chagas Disease. AIDS Patient Care STDs. 14:581-588.

Engel JC, Doyle PS, Hseih I, McKerrow JH. 1998. Cysteine protease inhibitors can cure experimental *Trypanosoma cruzi* infection. J Exp Med 188:725-734.

Higuchi M, Fukasawa S, Brito T, Parzianello LC, Belloti G, Ramires JA. 1999. Different microcirculatory and interstitial matrix patterns in idiopathic dilated cardiomyopathy and Chagas disease: a three dimensional confocal study. Heart 82:279-285.

Huang H, Calderon TM, Berman JW, Braunstein VL, Weiss LM, Wittner M, Tanowitz HB. 1999a. Infection of endothelial cells with *Trypanosoma cruzi* activates NF-κB and induces vascular adhesion molecule Expression. Infect Immun 67:5434-5440.

Huang H, Chan J, Wittner M, Jelicks LA, Braunstein VL, Bacchi CJ, Yarlett N, Chandra M, Shirani, J., Tanowitz, H.B. 1999b. Expression of cardiac cytokines and inducible form of nitric oxide synthase (NOS2) in *Trypanosoma cruzi*-infected mice. J Mol Cell Cardiol 31:75-88.

Huang H, Petkova SB, Pestell RG, Bouzahzah B, Chan J, Magazine H, Weiss LM, Christ GJ, Lisanti MP, Douglas SA, Shtutin V, Halonen SK, Wittner M, Tanowitz HB. 2000. *Trypanosoma cruzi* infection (Chagas disease) of mice causes activation of the mitogen-activated protein kinase cascade and expression of endothelin-1 in the myocardium. J Cardiovasc Pharmacol 36: (Suppl. 1) S148-150.

Lima APC, Almeida PC, Tersariol ILS, Schmitz V, Schmaier AH, Juliano L, Hirata IY, Müller-Esterl W, Chagas JR, Scharfstein J. 2002. Heparan sulfate modulates kinin-release by *Trypanosoma cruzi* through the activity of cruzipain. J Biol Chem 22:5875-5881.

Morrot A, Strickland DK, Higuchi Md, Reis M, Pedrosa R, Scharfstein J. 1997. Human T cell responses against the major cysteine proteinase (cruzipain) of *Trypanosoma cruzi*: role of the multifunctional alpha 2-macroglobulin receptor in antigen presentation by monocytes. Int Immunol 9:825-834.

Oliveira JS, De Oliveira M, Fredderigue U, Filho ECL. 1981. Apical aneurysm of Chagas heart disease. Br Heart J 46:432-437.

de Oliveira RA, de Moraes Silva, MA, de Andrade RR, Montenegro MRG. 2000. Clinical and pathological assessment of 82 patients with cardiovascular disease undergoing autopsy at the Hospital das Clinicas of the Faculdade de Medicina de Botucata from 1988-1993. Arq Bras Cardiol 75:308-312.

Parada H, Carrasco H, Anez N, Fuenmayor C, Inglessis I. 1997. Cardiac involvement is a constant finding in acute Chagas disease: a clinical, parasitological and histopathological study. Int J Cardiol 60:49-54.

Pereira-Barretto, A.C., Mady, C., Arteaga-Fernandez, E., Stolf, N., de Lourdes-Higuchi, M., Belloti, G., Pileggi, F. 1986. Right ventricular endomyocardial biopsy in chronic Chagas disease. Am Heart J. 111:307-310.

Petkova SB, Huang H, Factor SM, Pestell RG, Bouzahzah B, Jelicks LA, Weiss LM, Douglas SA, Wittner M, Tanowitz HB. 2001. Role of endothelin in the pathogenesis of Chagas disease. Int J Parasitol. 31:499-511.

Reis DD, Jones EM, Tostes S, Lopes ER, Chapadeiro E, Gazzinelli G, Colley DG, McCurley, TL. 1993. Expression of major histocompatibility complex antigens and adhesion molecules in hearts of patients with chronic Chagas disease. Am J Trop Med Hyg 49:192-200.

Rocha A, Ferreira MS, Nishioka SA, Silva AM, Burgarelli MK, Silva M, Moura LP, Ugrinovich R, Raffin CN. 1993. *Trypanosoma cruzi* meningoencephalitis and myocarditis in a patient with acquired immunodeficiency syndrome. Rev Inst Med Trop Sao Paulo 35:205-208.

Rossi MA, Souza AC. 1999. Is apoptosis a mechanism of cell death of cardiocytes in chronic chagasic myocarditis. Int J Cardiol. 325-331

Sartori AM, Shikanai-Yasuda MA, Amato Neto V, Lopes MH. 1998. Follow-up of 18 patients with human immunodeficiency virus infection and chronic Chagas disease, with reactivation of Chagas disease causing cardiac disease in three patients. Clin Infect Dis 26:177-179.

Sartori AM, Sotto MN, Braz LM, Oliveira Junior Oda C, Patzina RA, Barone AA, Shikanai-Yasuda MA. 1999. Reactivation of Chagas disease manifested by skin lesions in a patient with AIDS. Trans R Soc Trop Med Hyg 93:631-632.

Scharfstein J, Schmitz V, Morandi V, Capella MM, Lima AP, Morrot A, Juliano L, Muller-Esterl W. 2000. Host cell invasion by *Trypanosoma cruzi* is potentiated by activation of bradykinin B(2) receptors. J Exp Med 192:1289-1300.

Simoes MV, Pintya AO, Bromberg-Marin G, Sarabanda AV, Antloga CM, Pazin-Filho A, Maciel BC, Marin-Neto JA. 2000. Relation of regional sympathetic denervation and

Tanowitz et al.

myocardial perfusion disturbance to wall motion impairment in Chagas cardiomyopathy. Am J Cardiol 86:975-981.

Tanowitz, H.B., Kirchhoff, L.V., Simon, D., Morris, S.A., Weiss, L.M., Wittner, M. Chagas disease. 1992a. Clin Microbiol Rev 5:400-419.

Tanowitz HB, Gumprecht JP, Spurr D, Calderon TM, Ventura MC, Raventos-Suarez C, Factor SM, Hatcher V, Wittner M, Berman JW. 1992b. Cytokine gene expression of endothelial cells infected with *Trypanosoma cruzi*. J Infect Dis 166:598-603.

Zhang J, Andrade ZA, Yu Z-X, Andrade SG, Takeda K, Sadirgursky M, Ferrans VJ. 1999. Apoptosis in a canine model of acute chagasic myocarditis. J Mol Cell Cardiol 31:581-596.

THE CONTRIBUTION OF AUTOIMMUNITY TO CHAGAS HEART DISEASE

J. S. Leon and D. M. Engman
Departments of Microbiology-Immunology and Pathology
Feinberg Medical School of Northwestern University, Chicago, IL 60611

ABSTRACT

There are many potential mechanisms underlying the pathogenesis of chagasic heart disease. The frequent absence of parasites from the inflamed heart tissue of chronically infected individuals suggests that the disease may be, in part, autoimmune in nature. Mechanisms to explain the induction of *T. cruzi* induced autoimmunity include (i) polyclonal lymphocyte activation, induced by the parasite, (ii) bystander activation induced by tissue damage and stimulation of normally tolerant, self-reactive lymphocytes, (iii) altered antigen processing leading to the generation and presentation of "cryptic self epitopes," and (iv) molecular mimicry, immunity to a parasite epitope that cross-reacts with a self epitope that "mimics" it. The genetics of host and parasite also determine susceptibility to *T. cruzi* induced autoimmunity. To date, there is little evidence that the *T. cruzi* induced autoimmunity directly causes pathology in human Chagas disease or even in mouse models of the disease. Therefore, public health interventions should focus on control of the insects that transmit the parasite, development of parasiticidal drugs and vaccines, and testing of blood products since they are important sources of potential new infections.

INTRODUCTION

Chagas disease encompasses three main, largely non-coincident, pathologies: inflammation of the heart, dilation of the esophagus or colon, and abnormalities of the central nervous system, affecting roughly 30%, 5%, and less than 5% of individuals infected with protozoan parasite *Trypanosoma cruzi*, respectively (Moncayo, 1999) (see Chapter 10). These varied diseases typically develop years to decades after infection. Well over half of the infected individuals develop none of these sequelae. After nearly a century of investigation, the mechanisms of Chagas disease pathogenesis are unclear and under debate. Why only some individuals develop disease, why there is such variability in disease manifestations, why it takes so long for the disease to manifest, and what triggers disease initially are some of the many questions clinicians and researchers have investigated during the past several decades.

One proposed hypothesis is that Chagas disease is an autoimmune disease (reviewed in Eisen and Kahn, 1991; Kierszenbaum, 1999), an immune reaction against self antigens causing tissue inflammation or cellular damage (Abbas et al., 2000). Autoimmunity results when mechanisms responsible for mantaining immunological self-tolerance fail. Autoimmune responses may be present and measurable in the host in the absence of tissue inflammation or overt disease. Furthermore, the presence of antibodies or T cells reactive with host antigens may be a consequence and not a cause of tissue injury. To unequivocally prove that Chagas disease is an autoimmune disease, tissue inflammation or cellular damage must be shown to be directly caused by the autoimmune reactivity, whatever the initiating factor.

The clinical and public health significance of whether Chagas disease is an autoimmune disease is great. If Chagas disease is an autoimmune disease, then the therapies for the illness must address autoimmune mechanisms. Chemotherapies solely directed against the parasite might not prevent autoimmune tissue destruction; a two-pronged strategy to kill the parasite and reduce autoimmune tissue damage should be employed. In addition, potential anti-*T. cruzi* vaccine candidates would have to be screened to make sure that they do not induce an autoimmune disease. Thus, anti-*T. cruzi* chemotherapy and *T. cruzi* vaccines must be pursued with the potential for autoimmune sequelae in mind.

The hypothesis that Chagas disease is an autoimmune disease arose from initial studies on Chagas cardiac pathology and the discovery of *T. cruzi*-host cross-reactive antibodies. Histologic analysis of tissues from Chagas patients, particularly cardiac tissue, showed tissue inflammation and fibrosis occurring in the apparent absence of *T. cruzi*, suggesting that these inflammatory lesions were not initiated by *T. cruzi* or parasitized tissue as previously believed (Torres, 1941). Later, antibodies against host tissue were reported in Chagas patients (Cossio et al., 1974). This report was later retracted for methodological concerns (Khoury et al., 1983), but it encouraged other groups to search for and publish on autoantibodies in Chagas patients (reviewed in Kierszenbaum, 1986).

If inflammatory processes are not associated with *T. cruzi* or infected cells, and the immune system is targeting host tissue, what could initiate these inflammatory processes? In other words, is Chagas disease an autoimmune disease? The concept of Chagas disease as an autoimmune disease was put forth by Santos-Buch and Teixeira (Santos-Buch and Teixeira, 1974). In support of this possibility, several groups have reported that autoimmunity is induced upon infection with *T. cruzi* in humans and experimental animals and that this autoimmunity can directly contribute to pathology (discussed below and in Engman and Leon, 2002; Kierszenbaum, 1999). A second possibility is that inflammation is targeting "residual" *T. cruzi* antigens and that autoimmunity is an epiphenomenon. With the advent of sensitive nucleic acid based technologies, such as PCR and *in situ* hybridization, several groups have shown the presence of *T. cruzi* DNA in inflammatory foci and presumably protein antigen as well, even if intact amastigotes may not be present (reviewed by Tarleton, this volume). Therefore, evidence exists to support both hypotheses: (i) Chagas disease is an autoimmune disease and (ii) Chagas disease is caused by residual parasite and anti-parasite immunity. Several other potential mechanisms, including microvascular spasms, ischemia, and direct toxicity of the parasite will not be addressed here (Tanowitz et al., 1992, and this volume).

EVIDENCE THAT CHAGAS DISEASE IS AN AUTOIMMUNE DISEASE

The hypothesis that Chagas disease is an autoimmune disease should be considered as two separate issues: (i) autoimmunity is induced by infection with *T. cruzi*, and (ii) *T. cruzi* induced autoimmunity is pathogenic. The distinction between these two issues is, in our view, the most important source of confusion in the literature debating the autoimmune hypothesis for disease pathogenesis. Infection of humans with *T. cruzi* has been shown to induce autoantibodies against antigens in heart, skeletal muscle and nervous tissue (reviewed in Kierszenbaum, 1999). These include ribosomal P proteins (Levin et al., 1989), myosin (Cunha-Neto et al., 1995), β1 adrenoreceptor (Sterin-Borda and Borda, 2000), cytoskeletal microtubule associated proteins (Kerner et al., 1991), LIST neuronal proteins (Petry, 1989; Van Voorhis et al., 1991), and a novel mammalian protein, Cha (Girones et al., 2001). Many of these target proteins have ubiquitous expression that does not support the observation of heart specific inflammation in Chagas disease. Furthermore, there is little evidence that these antibodies are more prevalent in patients with Chagas disease than in asymptomatic, *T. cruzi* infected individuals. The clinical significance of these autoantibodies is not yet clear. T cells that recognize cardiac antigens have also been identified in chronically infected humans and mice. In humans, T cell clones from Chagas disease patients proliferated when cultured with a human cardiac myosin peptide or a parasite antigen, B13 (Cunha-Neto et al., 1996). In mice, T cells proliferated *in vitro* to heart homogenate (Ribeiro dos Santos et al., 1992) or to myosin (Rizzo et al., 1989).

To address whether *T. cruzi* induced autoimmunity is *pathogenic* researchers have studied the contribution of autoantibodies to disease. To date, there is no evidence that *T. cruzi* induced autoantibodies can induce disease upon transfer into a naïve host. However, *T. cruzi* induced antibodies do affect the contraction and cell signaling of cardiac myocytes *in vitro* (Borda and Sterin-Borda, 1996). In addition, sera from chronically infected mice lysed myocytes *in vitro* through an antibody dependent cytotoxicity mechanism (Laguens et al., 1988). Finally, immunization of mice with a *T. cruzi* antigen, cruzipain, induced autoantibodies to myosin, IgG deposits in heart sections, and conduction abnormalities in both the mother and pups (Giordanengo, 2000b). The authors of this research suggested that autoantibodies are pathogenic because of the association between the presence of autoantibodies and conduction abnormalities. Many of the targets of these autoantibodies are intracellular and thus it is difficult to understand how these autoantibodies cause disease if their target antigen(s) is inaccessible.

There is little evidence supporting the contribution of cellular autoimmunity to Chagas disease. One report demonstrated that splenocytes from a chronically infected mouse can lyse syngeneic myoblasts (Laguens et al., 1989). Transfer of splenocytes from infected mice stimulated *in vitro* with *T. cruzi* extract induced sciatic nerve (Hontebeyrie-Joskowicz et al., 1987) and cardiac inflammation (Laguens et al., 1989). The most compelling evidence supporting a role for cellular autoimmunity in disease pathogenesis was the finding that CD4[+] T cells from chronically infected mice mediated the rejection of implanted syngeneic newborn hearts (Ribeiro dos Santos et al., 1992). However, if a different combination of parasite and mouse strains was used, this did not occur (Tarleton et al., 1997). The conflicting nature of these results may be explained by differences in the ability of individual *T. cruzi* strains to induce autoimmunity or differences in the susceptibility of particular mouse strains to develop autoimmunity (discussed below). Recently, blocking immunity to heart antigens (tolerizing to heart antigens), enriched for myosin, reduced inflammation and fibrosis in *T. cruzi* infected mice (Pontes-De-Carvalho et al., 2002). Though this report offers compelling evidence for the pathogenicity of anti-heart antigen immunity in *T. cruzi* infected mice, there are methodological concerns which detract from its impact including: small sample size, no evidence that the immune system was tolerized to heart antigens, and discordant results of their positive control with published results (Godsel et al., 2001). Finally, immunization of mice with specific parasite proteins can induce autoimmunity and cardiac disease. Specifically, immunization of mice with regions of the *T. cruzi* ribosomal P1 and P2 protein induced production of autoantibodies to the mouse ribosomal proteins as well as electrocardiographic alterations (Motran et al., 2000). Immunization with cruzipain induced autoantibodies and T cell responses to myosin, skeletal myositis (Giordanengo et al., 2000a), and cardiac conduction abnormalities (Giordanengo et al., 2000b). Since no live parasites were used, the autoimmunity is believed to be induced through a molecular mimicry mechanism.

In conclusion, there is a large body of evidence that infection with   *T. cruzi* induces both humoral and cellular autoimmunity. However, the pathogenic potential of the autoimmunity has not been proven. *T. cruzi* induced autoantibodies affect cells *in vitro*, but have not been shown to have a direct role *in vivo*. Regarding cellular immunity, adoptive transfer and immunization experiments with parasite proteins do not necessarily recapitulate events in infected mice. As a result, no conclusion can be made about whether *T. cruzi*-induced autoimmunity directly contributes to tissue damage in human Chagas disease, or for that matter, in infected mice.

CRITICISMS OF THE AUTOIMMUNE DISEASE HYPOTHESIS

Two criticisms are levied against the autoimmune hypothesis of pathogenesis. One criticism is based on the fact that immunosuppressants, which generally relieve symptoms of autoimmune disease, exacerbate disease and mortality in Chagas patients. The best examples of this are Chagas heart transplant recipients receiving immunosuppressants, and Chagas patients infected with HIV. In both cases, the presence of the parasite confounds the question of whether autoimmunity contributes to disease, since suppressing host immunity results in an increased proliferation of parasites. For the record, the largest study on Chagas heart transplant recipients concluded that Chagas patients have no difference in mortality compared to heart transplant recipients suffering from idiopathic dilated cardiomyopathy or ischemic cardiomyopathy (Bocchi and Fiorelli, 2001). The second criticism posits that autoimmunity does not contribute to disease because *T. cruzi* chemotherapy alone reduces clinical disease in humans and experimental animals. However, there is no consensus on the efficacy of chemotherapy on human disease (Viotti et al., 1994; Parada et al., 1997; Bahia-Oliveira et al., 2000; Fabbro De Suasnabar et al., 2000; Inglessis et al., 1998; Lauria-Pires et al., 2000). Unless chemotherapy completely eliminates disease, any residual disease can be explained by additional mechanisms. In experimental models of *T. cruzi* infection, chemotherapy given *immediately* after infection reduces and sometimes eliminates cardiac disease (Urbina, 1999). Because *T. cruzi* is the trigger for autoimmunity, elimination of this trigger in the *acute* disease phase could potentially eliminate the induction of autoimmunity, making the analysis of the contribution of autoimmunity irrelevant.

POTENTIAL MECHANISMS OF *T. CRUZI*-INDUCED AUTOIMMUNITY

How can *T. cruzi* induce autoimmunity in a host which is normally tolerant to its own antigens? Four possible explanations include the mechanisms of polyclonal activation, bystander activation, cryptic epitope, and molecular mimicry (reviewed in Leon and Engman, 2001).

Polyclonal activation. Autoimmunity may result from antigen-independent stimulation of self-reactive lymphocytes that are not deleted during lymph development. Polyclonal activators stimulate many T and B lymphocytes, irrespective of antigen specificity and, in some cases, by interacting with surface molecules other than antigen receptors. The most common example of a polyclonal activator is lipopolysaccharide, which induces a wide repertoire of acute autoantibodies in mice (Granholm and Cavallo, 1992). These autoantibodies have weak affinities and are often of the IgM isotype. Certain *T. cruzi* strains have been shown to possess polyclonal activators, suggesting that polyclonal activation may be a possible mechanism for the induction of *T. cruzi*-induced autoimmunity (Minoprio, 2001).

Bystander activation. In the mechanism of bystander activation, *T. cruzi* infection causes tissue destruction and release of host antigens. The excess levels of released host antigens in the presence of an environment rich in proinflammatory cytokines, nitric oxide and chemokines may overcome self-tolerance and initiate autoimmunity. Evidence for this hypothesis includes the observation that cardiac autoimmunity can result from many, varied types of insults to the heart, including transplantation, surgery, and infection (Neu et al., 1987; de Scheerder et al., 1989; Fedoseyeva et al., 1999).

Cryptic epitope. The third mechanism, cryptic epitope, suggests that either (i) *T. cruzi* infection releases previously sequestered epitopes or (ii) that the inflammatory environment induced by *T. cruzi* induces the processing and immune presentation of novel self epitopes. Immunity against these novel epitopes is rapidly induced because the immune system is not tolerant to these novel epitopes (Lanzavecchia, 1995). In support of this hypothesis, antigen processing and presentation is altered after *in vitro* treatment with IFN-γ (York et al., 1999). The mechanism of cryptic epitope has been used to explain the genesis of rheumatoid arthritis and systemic lupus erythematosus (Warnock and Goodacre, 1997).

Molecular mimicry. The last and arguably most popular mechanism is molecular mimicry. Molecular mimicry leads to autoimmunity as a result of a "misdirected" immune response. When a parasite antigen closely resembles a host antigen, the immune system may be induced first against the parasite antigen and then "cross-react" with a self antigen, causing autoimmunity. Evidence for this hypothesis includes the reports of autoimmunity upon immunization with a *T. cruzi* antigen (Giordanengo et al., 2000a; Motran et al., 2000) and the many reports of cross-reactive (*T. cruzi*-host) autoantibodies such as B13-myosin (Cunha-Neto et al., 1995), or cruzipain-myosin (Giordanengo et al., 2000b), *T. cruzi*-mammalian ribosomal P proteins (Motran et al., 2000) and *T. cruzi* shed acute phase antigen-Cha autoantigen (Girones et al., 2001).

Epitope spreading. As a final twist on the bystander activation mechanism, the autoantigen that initiates autoimmunity may not be the autoantigen involved during development of disease. This phenomenon is called "epitope spreading" and it describes how the primary epitope/antigen target of autoimmunity may change during the course of disease. At one point, autoimmunity against one epitope develops, causing damage to tissue(s) containing that epitope. This damage then results in the release of additional self antigens, the processing and presenting of which induces the stimulation of autoimmunity of additional epitope specificity(ies). Interestingly, the initial responses typically wane as immunoregulatory mechanisms kick in, leading to "waves" of autoimmunity targeted to different epitopes/antigens (Vanderlugt and Miller, 1996).

When is autoimmunity first induced by *T. cruzi*? Autoimmunity may be induced immediately after the initial contact of the parasite with the host, during the acute phase of disease (Ternynck et al., 1990; Grauert et al., 1993; Leon et al., 2001). This early autoimmunity likely results from tissue damage caused by the parasite and/or molecular mimicry. The polyclonal, polyspecific nature of the autoantibody response supports the former hypothesis. Autoimmunity may also develop later in the disease course (Acosta and Santos-Buch, 1985; Laguens et al., 1988). Persistent, chronic inflammation may be necessary to overcome the threshold of cardiac damage or produce the correct inflammatory environment for the stimulation and expansion of autoreactive cells. In closing, these mechanisms are not exclusive of each other and combinations of these mechanisms may play a role in the induction of autoimmunity during Chagas disease.

Host immunogenetics and parasite genetics in autoimmunity

Immunogenetic factors of the host and genetic characteristics of the parasite may also influence the induction of autoimmunity. Factors that may influence the induction of autoimmunity in the host include genetic background, gender, and age among others. To date, the role of host genetic background is the only factor examined in the induction of autoimmunity caused by *T. cruzi* infection. Certain strains of mice are susceptible while others are resistant. Acute infection of A/J mice with *T. cruzi* induces cellular and humoral autoimmunity against myosin while acute infection of C57BL/6 mice does not (Leon et al., 2001). There is evidence that *T. cruzi* genetics determine the induction of autoimmunity. Infection of mice with the Brazil strain of *T. cruzi* induces anti-heart antibodies whereas infection with the Guayas strain of *T. cruzi* does not (Tibbetts et al., 1994; Rowland et al., 1995).

CONCLUSIONS

In conclusion, there is clear evidence that autoimmunity can be induced by *T. cruzi* infection in humans and experimental animals (Figure 1). However, there is no proof that autoimmunity directly contributes to the pathogenesis of Chagas disease. It should also be emphasized that it is presumed by many (see Tarleton chapter) but not proven that *in vivo* anti-*T. cruzi* immunity is responsible for inflammation and damage present in the myocardium.

It may be that the cause of Chagas disease is autoimmunity, *T. cruzi* and its associated anti-*T. cruzi* responses, other mechanisms, or a combination. If both autoimmunity and anti-*T. cruzi* immune responses contribute to Chagas disease, then it is necessary to address the relative contributions of both to disease in human patients. The balance in certain subpopulations may be skewed more towards the parasite than autoimmunity and vice versa. If autoimmunity plays no part in disease, then how is it induced and will it disappear once the parasite is cleared? Various mechanisms may explain how an infectious organism can break immunologic self tolerance. To convince skeptics that *T. cruzi*-induced autoimmunity contributes to the pathology of Chagas disease additional evidence is required. Therefore, from the public health perspective, anti-*T. cruzi* chemotherapy and vaccine trials are worth pursuing until Chagas disease is proved to have an autoimmune component to pathogenesis.

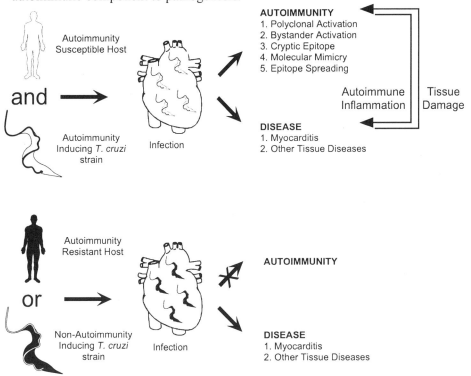

Figure 1. Model of autoimmunity induction in Chagas heart disease. Host infection with *T. cruzi* may induce autoimmunity and/or disease depending on host immunogenetics and parasite genetics. Autoimmunity may be induced via a sole or a combination of mechanisms listed under Autoimmunity. Autoimmunity may progress to disease (Autoimmune Inflammation) and disease may induce autoimmunity through tissue damage (Tissue Damage).

References

Abbas AK, Lichtman AH, Pober JS. 2000. Cellular and Molecular Immunology. W B Saunders Co, Philadelphia, pp. 553.

Acosta AM, Santos-Buch CA. 1985. Autoimmune myocarditis induced by *Trypanosoma cruzi*. Circulation 71:1255-61.

Bahia-Oliveira LM, Gomes JA, Cancado JR, Ferrari TC, Lemos EM, Luz ZM, Moreira MC, Gazzinelli G, Correa-Oliveira R. 2000. Immunological and clinical evaluation of chagasic patients subjected to chemotherapy during the acute phase of *Trypanosoma cruzi* infection 14-30 years ago. J Infect Dis 182:634-8.

Bocchi EA, Fiorelli A. 2001. The paradox of survival results after heart transplantation for cardiomyopathy caused by *Trypanosoma cruzi*. First Guidelines Group for Heart Transplantation of the Brazilian Society of Cardiology. Ann Thorac Surg 71:1833-8.

Borda ES, Sterin-Borda L. 1996. Antiadrenergic and muscarinic receptor antibodies in Chagas' cardiomyopathy. Int J Cardiol 54:149-56.

Cossio PM, Diez C, Szarfman A, Kreutzer E, Candiolo B, Arana RM. 1974. Chagasic cardiopathy. Demonstration of a serum gamma globulin factor which reacts with endocardium and vascular structures. Circulation 49:13-21.

Cunha-Neto E, Duranti M, Gruber A, Zingales B, de Messias I, Stolf N, Bellotti G, Patarroyo, ME, Pilleggi F, Kalil J. 1995. Autoimmunity in Chagas' disease cardiomyopathy: biological relevance of a cardiac myosin-specific epitope cross-reactive to an immunodominant *Trypanosoma cruzi* antigen. Proc Natl Acad Sci USA 92:3541-45.

Cunha-Neto E, Coelho V, Guilherme L, Fiorelli A, Stolf N, Kalil J. 1996. Autoimmunity in Chagas' disease: identification of cardiac myosin-B13 *Trypanosoma cruzi* protein crossreactive T cell clones in heart lesions of a chronic Chagas' cardiomyopathy patient. J Clin Invest 98:1709-12.

de Scheerder IK, de Buyzere ML, Delanghe JR, Clement DL, Wieme RJ. 1989. Anti-myosin humoral immune response following cardiac injury. Autoimmunity 4:51-8.

Eisen H, Kahn S. 1991. Mimicry in *Trypanosoma cruzi*: fantasy and reality. Curr Opin Immunol 3:507-10.

Engman DM, Leon JS. 2002. Pathogenesis of Chagas heart disease: role of autoimmunity. Acta Trop 81:123-32.

Fabbro De Suasnabar D, Arias E, Streiger M, Piacenza M, Ingaramo M, Del Barco M, Amicone N. 2000. Evolutive behavior towards cardiomyopathy of treated (nifurtimox or benznidazole) and untreated chronic chagasic patients. Rev Inst Med Trop Sao Paulo 42:99-109.

Fedoseyeva EV, Zhang F, Orr PL, Levin D, Buncke HJ, Benichou G. 1999. De novo autoimmunity to cardiac myosin after heart transplantation and its contribution to the rejection process. J Immunol 162:6836-42.

Godsel LM, Wang K, Schodin BA, Leon JS, Miller SD, Engman DM. 2001. Prevention of autoimmune myocarditis through the induction of antigen-specific peripheral immune tolerance. Circulation 103:1709-14.

Giordanengo L, Fretes R, Diaz H, Cano R, Bacile A, Vottero-Cima E, Gea S. 2000a. Cruzipain induces autoimmune response against skeletal muscle and tissue damage in mice. Muscle Nerve 23:1407-1413.

Giordanengo L, Maldonado C, Rivarola HW, Iosa D, Girones N, Fresno M, Gea S. 2000b. Induction of antibodies reactive to cardiac myosin and development of heart alterations in cruzipain-immunized mice and their offspring. Eur J Immunol 30:3181-9.

Girones N, Rodriguez CI, Carrasco-Marin E, Hernaez RF, de Rego JL, Fresno M. 2001. Dominant T- and B-cell epitopes in an autoantigen linked to Chagas' disease. J Clin Invest 107:985-93.

Granholm NA, Cavallo T. 1992. Autoimmunity, polyclonal B-cell activation and infection. Lupus 1:63-74.

Grauert MR, Houdayer M, Hontebeyrie-Joskowciz M. 1993. *Trypanosoma cruzi* infection enhances polyreactive antibody response in an acute case of human Chagas' disease. Clin Exp Immunol 93:85-92.

Hontebeyrie-Joskowicz M, Said G, Milon G, Marchal G, Eisen H. 1987. L3T4[+] T cells able to mediate parasite-specific delayed-type hypersensitivity play a role in the pathology of experimental Chagas' disease. Eur J Immunol 17:1027-33.

Inglessis I., Carrasco HA, Anez N, Fuenmayor C, Parada H, Pacheco JA, Carrasco HR. 1998. Clinical, parasitological and histopathologic follow-up studies of acute Chagas patients treated with benznidazole. Arch Inst Cardiol Mex 68:405-10.

Khoury EL, Diez C, Cossio PM, Arana RM. 1983. Heterophil nature of EVI antibody in *Trypanosoma cruzi* infection. Clin Immunol Immunopathol 27:283-8.

Kierszenbaum F. 1986. Autoimmunity in Chagas' disease. J Parasitol 72:201-211.

Kierszenbaum F. 1999. Chagas' disease and the autoimmunity hypothesis. Clin Microbiol Rev 12:210-23.

Laguens RP, Cabeza Meckert PM, Chambo JG. 1989. Immunologic studies on a murine model of Chagas disease. Medicina (Buenos Aires) 49:197-202.

Laguens RP, Meckert PC, Chambo JG. 1988. Antiheart antibody-dependent cytotoxicity in the sera from mice chronically infected with *Trypanosoma cruzi*. Infect Immun. 56:993-7.

Lanzavecchia A. 1995. How can cryptic epitopes trigger autoimmunity? J Exp Med 181:1945-8.

Lauria-Pires L, Braga MS, Vexenat AC, Nitz, N, Simoes-Barbosa A, Tinoco DL, Teixeira AR. 2000. Progressive chronic Chagas heart disease ten years after treatment with anti-*Trypanosoma cruzi* nitroderivatives. Am J Trop Med Hyg 63:111-8.

Leon JS, Engman DM. 2001. Autoimmunity in Chagas heart disease. Int J Parasitol. 31:554-60.

Leon JS, Godsel LM, Wang K, Engman DM. 2001. Cardiac myosin autoimmunity in acute Chagas heart disease. Infect Immun 69:5643-9.

Levin MJ, Mesri E, Benarous R, Levitus G, Schijman A, Levy-Yeyati P, Chiale PA, Ruiz AM, Kahn A, Rosenbaum MB, *et al.* 1989. Identification of major *Trypanosoma cruzi* antigenic determinants in chronic Chagas' heart disease. Am J Trop Med Hyg 41:530-8.

Kerner N, Liegeard P, Levin MJ, Hontebeyrie-Joskowicz M. 1991. *Trypanosoma cruzi*: antibodies to a MAP-like protein in chronic Chagas' disease cross-react with mammalian cytoskeleton. Exp Parasitol 73:451-9.

Minoprio P. 2001. Parasite polyclonal activators: new targets for vaccination approaches? Int J Parasitol 31:588-91.

Moncayo A. 1999. Progress towards interruption of transmission of Chagas disease. Mem Inst Oswaldo Cruz 94 Suppl 1:401-4.

Motran CC, Fretes RE, Cerban FM, Rivarola HW, Vottero de Cima E. 2000. Immunization with the C-terminal region of *Trypanosoma cruzi* ribosomal P1 and P2 proteins induces long-term duration cross-reactive antibodies with heart functional and structural alterations in young and aged mice. Clin Immunol 97:89-94.

Neu N, Craig SW, Rose NR, Alvarez F, Beisel KW. 1987. Coxsackievirus induced myocarditis in mice: cardiac myosin autoantibodies do not cross-react with the virus. Clin Exp Immunol 69:566-74.

Parada H, Carrasco HA, Anez N, Fuenmayor C, Inglessis I. 1997. Cardiac involvement is a constant finding in acute Chagas' disease: a clinical, parasitological and histopathological study. Int J Cardiol 60:49-54.

Petry K, Eisen H. 1989. Chagas' disease: a model for the study of autoimmune diseases. Parasitol Today 5:111-121.

Pontes de Carvalho, L, Santana CC, Soares MB, Oliveira GG, Cunha-Neto E, Ribeiro-Dos-Santos R. 2002. Experimental chronic Chagas' disease myocarditis is an autoimmune disease preventable by induction of immunological tolerance to myocardial antigens. J Autoimmun 18:131-8.

Ribeiro dos Santos R, Rossi MA, Laus JL, Silva JS, Silvino W, Mengels J. 1992. Anti-CD4 abrogates rejection and reestablishes long-term tolerance to syngeneic newborn hearts grafted in mice chronically infected with *Trypanosoma cruzi*. J Exp Med 175:29-39.

Rowland E, Luo H, McCormick T. 1995. Infection characteristics of an Ecuadorian *Trypanosoma cruzi* strain with reduced virulence. J Parasitol 81:123-6.

Rizzo LV, Cunha-Neto E, Teixeira AR. 1989. Autoimmunity in Chagas' disease: specific inhibition of reactivity of CD4$^+$ T cells against myosin in mice chronically infected with *Trypanosoma cruzi*. Infect Immun 57:2640-4.

Santos-Buch CA, Teixeira AR. 1974. The immunology of experimental Chagas' disease. 3. Rejection of allogeneic heart cells in vitro. J Exp Med 140:38-53.

Sterin-Borda L, Borda E. 2000. Role of neurotransmitter autoantibodies in the pathogenesis of chagasic peripheral dysautonomia. Ann N Y Acad Sci 917:273-80.

Tanowitz HB, Kirchhoff LV, Simon D, Morris SA, Weiss LM, Wittner M. 1992. Chagas' disease. Clin Microbiol Rev 5:400-19.

Tarleton RL, Zhang L, Downs MO. 1997. "Autoimmune rejection" of neonatal heart transplants in experimental Chagas disease is a parasite-specific response to infected host tissue. Proc Natl Acad Sci USA 94:3932-7.

Ternynck T, Bleux C, Gregoire J, Avrameas S, Kanellopoulos-Langevin C. 1990. Comparison between autoantibodies arising during *Trypanosoma cruzi* infection in mice and natural autoantibodies. J Immunol 144:1504-1511.

Tibbetts RS, McCormick TS, Rowland EC, Miller SD, Engman DM. 1994. Cardiac antigen-specific autoantibody production is associated with cardiomyopathy in *Trypanosoma cruzi*-infected mice. J Immunol 152:1493-9.

Torres CM. 1941. Sobre a anatomia patologica da doenca de Chagas Mem Inst Oswaldo Cruz 36:391-404.

Urbina JA. 1999. Chemotherapy of Chagas' disease: the how and the why. J Mol Med 77:332-8.

Vanderlugt CJ, Miller SD. 1996. Epitope spreading. Curr Opin Immunol 8:831-6.

Van Voorhis WC, Schlekewy L, Trong HL. 1991. Molecular mimicry by *Trypanosoma cruzi*: the Fl-160 epitope that mimics mammalian nerve can be mapped to a 12-amino acid peptide. Proc Natl Acad Sci USA 88:5993-5997.

Viotti R, Vigliano C, Armenti H, Segura E. 1994. Treatment of chronic Chagas' disease with benznidazole: clinical and serologic evolution of patients with long-term follow-up. Am Heart J 127:151-62.

Warnock MG, Goodacre JA. 1997. Cryptic T-cell epitopes and their role in the pathogenesis of autoimmune diseases. Br J Rheumatol 36:1144-50.

York IA, Goldberg AL, Mo XY, Rock KL. 1999. Proteolysis and class I major histocompatibility complex antigen presentation. Immunol Rev 172:49-66.

TRYPANOSOMA CRUZI AND CHAGAS DISEASE: CAUSE AND EFFECT

R. L. Tarleton
Center for Tropical and Emerging Global Diseases
The University of Georgia, Athens, GA 30602

ABSTRACT

Despite the fact that the association between infection with *T. cruzi* and the disease known as Chagas disease was discovered nearly 100 years ago, the cause and effect relationship between the infection and the severity of disease is still hotly debated. The crux of the debate revolves around the question of what instigates and perpetuates the damage that ultimately results in disease symptoms: the parasite and the immune response to it, or the host, as a result of failure to regulate autoreactive immune responses. It is generally (although not universally) agreed that in both cases it is the host immune response, and not direct parasite-mediated destruction of tissue, which is responsible for tissue damage. Much of the literature in this area tends to treat *T. cruzi* infection and Chagas disease as two almost completely separate, even unrelated, entities. This chapter presents arguments in favor of a tight association between the course of *T. cruzi* infection, and in particular the persistence of *T. cruzi* in chronically infected hosts, and the occurrence and severity of Chagas disease.

INTRODUCTION

Infection with *T. cruzi* results in a life-long relationship between parasite and host; no firm documentation of spontaneous and complete clearance of the infection exists. In the case of humans, initial infection often occurs in children or young adults but clinical disease is not evident until decades later. The acute stage of the infection is characterized by relatively high blood and tissue parasite levels and in the distribution of parasites to nearly all tissues and organs. These characteristics are the result of successful evasion of the innate immune response and the ability of *T. cruzi* to infect a wide range of host cell types. Although the acute phase is sometimes associated with severe symptoms, death from acute infection is relatively rare and generally occurs only in young children or those with primary or secondary immunodeficiencies.

The acute phase ends when the development of potent and largely effective immune responses controls the blood and tissue parasite levels, yielding a chronic persistent infection. This post-acute period is often separated into "indeterminate" (asymptomatic) and "chronic" (symptomatic) phases. However, it is as likely that these two "phases" are part of a continuum during which the damage that accumulates in the asymptomatic period eventually results in the clinical symptoms characteristic of chronic Chagas disease. The rate of conversion from asymptomatic to symptomatic chronic disease has been estimated to be approximately 1% per year. Although symptomatic disease occurs in only a minority (30-40%) of infected individuals, the determinants of the presence or absence of symptomatic disease, as discussed further below, are not clear. In the majority of cases, disease is most evident in the heart and, in a lower proportion of patients, in the digestive tract. Skeletal muscle disease is also obvious in experimental models although this has not been widely studied in human infections (Laguens et al., 1975; Sanz et al., 1978).

IMMUNITY TO *T. CRUZI*

Immune control of *T. cruzi* requires the combined response of a number of immune effector mechanisms. At a minimum, the generation of a substantial antibody response, and the activation of both $CD4^+$ and $CD8^+$ T cell compartments is needed to prevent death from an overwhelming acute parasitemia (Tarleton et al., 1996; Tarleton, 1997; Kumar and Tarleton, 1998). Additionally, these same responses are probably required to maintain control of *T. cruzi* during the chronic phase of the infection. The evidence for the latter comes from experimental models and natural human infections in which targeted depletions, immunosuppressive treatments or infection-induced immunosuppression result in an exacerbation of the infection (Jardim and Takayanagui, 1994; Tarleton et al., 1994; Sartori et al., 1995; Almeida et al., 1996; Tarleton et al., 1996). Cumulatively these results strongly suggest that there is little difference between the immune mechanisms that initially control the acute infection and the mechanisms maintaining this control in the chronic phase.

Qualitative aspects of the host immune response also influence the course and severity of the infection and disease. For example, it has become clear from recent studies that a type 1 biased T cell response (Th1/Tc1 with production of the type 1 cytokines IFN-gamma and IL-2) is protective, while a type 2-biased response (characterized by IL-4, IL-5 and IL-13 production), or a balanced type1/type2 response, is associated with more severe infection and disease (Barbosa de Oliveira et al., 1996; Hoft et al., 2000; Tarleton et al., 2000; Kumar and Tarleton 2001). Likewise, both qualitative and quantitative aspects of the B cell response, including the preferential antibody subclasses produced and the antigenic targets of these antibodies, are likely determinants of the success in immune control during *T. cruzi* infection.

THE RELATIONSHIP BETWEEN PARASITE PERSISTENCE AND THE PRESENCE/SEVERITY OF CHAGAS DISEASE

In the vast majority of natural, immunocompetent hosts, including most humans, the immune responses generated during the acute phase of *T. cruzi* infection control the parasite burden. However, even in the best of circumstances, with the generation of potent and appropriately focused immune responses, *T. cruzi* nevertheless manages to persist and is never completely cleared. In chronically infected hosts, the detection of parasites or parasite products in the blood as well as in muscle tissue often requires multiple sampling and highly sensitive means (e.g. PCR, *in situ* PCR, and immunohistochemistry). But there is no doubt that *T. cruzi* persists in infected individuals. Furthermore, there is strong evidence that the presence and level of circulating parasites during the acute and chronic stages of the infection correlates well with the severity of chronic disease. For example, initiating infections in mice using increasing numbers of parasites results in more severe disease 300 to 500 days later (Marinho et al., 1999). Also, human patients with cardiomyopathy have a higher prevalence of parasitemia than asymptomatic individuals (Salomone et al., 2000).

Chronic *T. cruzi* infection results in tissue-specific disease focused primarily in the gut and heart. If disease severity is linked to or even determined by parasite persistence, then one would expect that *T. cruzi* would be more evident in the tissues which are the focus of clinical disease. Numerous studies now provide evidence that this is exactly the case. Tissue-specific control of *T. cruzi* infection is evident early in the acute infection (Postan et al., 1983; Monteon et al., 1996) and is even more obvious in the chronic stage (Jones et al., 1992; Vago et al., 1996b; Zhang and Tarleton, 1999; Silva 2001), with persistence most commonly restricted to muscle and, to a lesser extent, the central nervous system. It is not at all clear what factors allow for parasite clearance from some tissues but not from others. Parasite strain-specific factors almost certainly contribute (Vago et al., 1996a; Andrade et al., 1999; Camargos et al., 2000; Vago et al., 2000). But it is unlikely that tissue-restriction is due solely to differential tropism of distinct strains since parasites are widely distributed in most if not all tissues and organs early in the infection, before the full development of B and T cell responses. More likely, the tissue-specific persistence of *T. cruzi* in the chronic infection is also related to qualitative and quantitative aspects of the local immune responses in these tissues.

A wealth of data from both experimental models and human infections provide solid evidence that the tissue-specific persistence of *T. cruzi* is coincident with the presence of inflammatory disease in these tissues and with the resulting clinical illness. A variety of techniques, including immunohistochemistry, whole tissue and *in situ* PCR and in situ hybridization have been used to unequivocally demonstrate the persistence of *T. cruzi* at sites of disease (reviewed in Tarleton and Zhang, 1999). Furthermore, intact amastigotes have been detected by careful and thorough investigation of heart tissue chagasic patients (Anez et al., 1999; Palomino et al., 2000). Perhaps the most compelling data documenting the link between tissue-specific persistence of *T. cruzi* and clinical disease comes from work demonstrating the presence of parasite DNA in the heart or esophageal tissue of individuals with cardiac or esophageal disease, respectively, but not in other infected individuals who lacked signs of disease in these tissues (Jones et al., 1993; Vago et al., 1996b). Similar results documenting the absolute association between the presence of parasites and the presence of disease in specific host tissues have also been presented in mouse models of the chronic infection (Zhang and Tarleton, 1999).

The coincidence of parasite persistence in specific tissues and the presence of tissue-specific disease in chronic *T. cruzi* infections would not come as a great surprise were it not for the many studies suggesting that parasites are essentially absent at sites of disease. And it was primarily these observations, based on standard histological techniques, which prompted the development of alternative explanations for the severity and the localization of disease (Kalil and Cunha-Neto, 1996). These explanations ultimately led to the dominance of the hypothesis that it is the immune response against host tissues that is the cause of Chagas disease. However, with the data provided by the application of sensitive methods for detection of parasites and parasite products in tissues, the need for an alternative explanation for tissue-specific disease in chronic *T. cruzi* infection no longer exists.

THE EVOLUTION OF DISEASE IN CHRONIC *T. CRUZI* INFECTION

One of the major difficulties in trying to understand the evolution of Chagas disease is that we generally are looking at a static picture provided by histological sections and are attempting to determine how the disease developed over time. Unfortunately, we do not yet have the tools to follow a particular lesion from its origins to its resolution. And without such tools, the same picture or tissue section can be interpreted in multiple ways. For example, one investigator may look at a site of inflammation that lacks readily detectable parasites and conclude that the inflammation is the result of immune responses directed against something other than *T. cruzi* (e.g. see Palomino et al., 2000). The most obvious candidates are self-antigens and the conclusion is that the lesion represents a failure of the immune system to regulate anti-self reactions. Alternatively another investigator can look at the same sites and see it as a success of the immune response - the infected cell or cells that were the likely initiators of the response have been detected and destroyed.

In the case of *T. cruzi* infection, the interpretation of the second investigator seems more plausible for a number of reasons. First we know the parasites are present, and that they are largely restricted to the tissues where disease is most evident. Second, intact parasites, parasite-infected cells, parasite antigens and parasite DNA are all detectable within the lesions themselves (Tarleton and Zhang, 1999). If the lesions are the result of autoimmune responses, why are these responses so focal and restricted only to areas of parasite persistence - what is different about the muscle cells in the middle of an inflammatory focus and those a few microns away which are not being "attacked"? Third, careful examination of affected tissues reveals the full range of evolving foci, starting with free parasites to cells infected with a few or many parasites to the presence of parasite antigen but not intact parasites in an inflammatory focus to eventual fibrosis and scarring (Palomino, 2000). Looking at this full continuum of lesions at different stages or resolution, one can start to envision how the disease is evolving.

Thus, while it is true that intact parasites are absent from many mature inflammatory foci the most likely explanation for this observation is that the inflammatory response has successfully destroyed the parasites in such foci. If the destruction of parasites and parasite-infected cells is considered a "successful" response, then why is tissue damage so extensive and why is clinical disease so severe? Is this a case of an over-aggressive immune response? Cellular immune responses almost always result in peripheral damage and this is also the case with *T. cruzi* infection; destruction of infected cells leaves behind dead and dying cells and scar tissue. Certainly, too much of a good thing is not good in itself. The best direct example of such an over-reaction is observed in mice that lack the IL-10 gene. These animals succumb to an overwhelming inflammatory disease early in infection with *T. cruzi* (Hunter et al., 1997). However, this result seems to be a rather unique case. It is perhaps, a tribute to the critical importance of IL-10 as a moderator of immune responses in general, rather than evidence that human Chagas disease is a result of a similar over-reaction.

As discussed above the general consensus is that type 1 biased responses are protective in *T. cruzi* infection while type 2- biased and mixed type 1/type 2 responses are exacerbative. In all cases where type 1 responses are potentiated, animals have lower parasite load and less severe disease (Barbosa de Oliveira et al., 1996; Hoft et al., 2000; Tarleton et al., 2000; Chen et al., 2001; Kumar and Tarleton 2001). Again an example of *T. cruzi* infection in a mouse knock-out strain is instructive (Tarleton et al., 2000). Mice lacking the *Stat4* gene as a result of targeted disruption fail to signal through the IL-12 receptor and as a result are deficient in the generation of type 1 cytokine responses. As expected, these animals also fail to control the acute phase of *T. cruzi* infection and succumb with high blood and tissue parasite levels. In sharp contrast, mice lacking *Stat6* gene function and which are as a result, deficient in the generation of type 2 cytokine responses, are at least as resistant as wild-type mice to infection with *T. cruzi*. In fact, the *Stat6* knockout mice show lower tissue parasite burdens and less evidence of chronic disease than similarly infected wild-type mice. The presence of potent type 1 cytokine responses with insufficient, down-regulatory type 2 responses have been proposed as the cause of a number of disorders classified as autoimmune, including Chagas disease (Rose, 1998). However, the contrary is clearly true in *T. cruzi* infection - type 2 cytokine responses not only are not required to down-regulate the type 1 response, the presence of type 2 responses actually exacerbates the infection and the disease.

The results of studies in the *Stat* knockouts, as well as other knockout strains of mice all suggest that parasite load in the acute and chronic stages of infection and the severity of disease are determined primarily, if not exclusively by the efficiency of the anti-parasite immune response. These results are further supported by data in human patients where it is clear that suppression of immune responses results in increased disease severity. If anti-parasite responses are appropriately focused, sufficiently strong and of the correct composition, then *T. cruzi* infection may be controllable and disease development prevented. However if these criteria are not met, then the parasite load will be higher, inflammatory foci will be more numerous and larger, and disease onset will be more rapid and more severe.

A MODEL FOR DISEASE DEVELOPMENT IN CHRONIC *T. CRUZI* INFECTION

If the success of *T. cruzi* infection and the severity of disease are tightly linked as is proposed here, how does one explain the variation in disease in *T. cruzi*-infected individuals? All chronically infected individuals have a persistent parasite burden but only a proportion of these individuals develops clinical disease. The figure below shows a proposed model for disease development in *T. cruzi* infection, based upon the data and conclusions discussed above. Looking first at parasite load, data from all experimental models suggest that parasite numbers in both the blood and in tissues is highest during the acute phase of the infection and drops substantially once an effective adaptive immune response is generated. Minimally this response includes appropriate antibody, and $CD4^+$ (helper) and $CD8^+$ (cytotoxic) T cell responses. The vast majority of individuals generate effective responses and thus control the infection. But failure to control the infection at this point, or the loss of the ability to maintain these responses at later points in the infection, results in high parasite loads and high risk of death (not shown).

In the acute as well as the chronic stage of the infection, it is likely that a range of parasite loads exist in different individuals although there is not a great deal of experimental data documenting this, nor indicating the extremes of the range. Even less is known about how variable the parasite load is over time in chronically infected individuals. However the available evidence from both humans and experimental animals indicates that except for cases of immunosuppression, the tissue and blood parasite levels in the chronic phase are orders of magnitude lower than those of the acute phase.

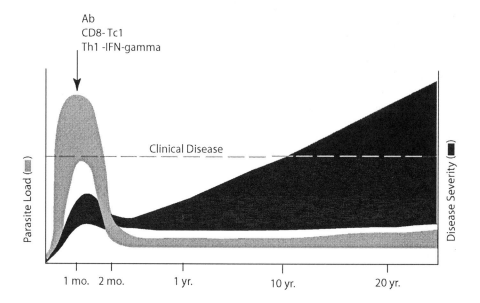

The important points of this model with respect to disease are that there is a wide variation in the intensity of disease among a population of infected individuals and that disease is essentially a cumulative phenomenon. Clinical illness occurs only in the most severely affected individuals. Well-studied experimental models suggest that the high tissue parasite load and the inflammatory responses that these parasites evoke are the major causes of tissue damage in the acute infection. As the parasite load drops with the generation of a controlling immune response, the severity of disease also decreases or at least levels off. But in nearly all individuals, parasite persistence guarantees that tissue destruction at some level will continue for the remainder of life. For the 60-70% of individuals who never develop clinical symptoms, the severity of disease may not rise above the baseline level established in the acute phase, or will rise only modestly. For the remainder of individuals, the effects of persistent infection and reaction to the infection cumulate in clinical disease as a threshold level of tissue damage is surpassed.

There are at least three explanations for why disease development and disease severity differs among individuals with *T. cruzi* infection:
1) differences in parasite load or the infecting strain
2) differential development of anti-self immune responses (i.e. autoimmunity)
3) differential ability to efficiently control *T. cruzi* without peripheral tissue damage.

Parasite strain and load differ between individuals and the characteristics of the infecting strain(s) of *T. cruzi* and the level at which these parasites persist in the chronic infection are certainly factors in disease severity. However given the relatively low parasite levels in all chronically infected patients, it seems more likely that parasite variation is a contributor to rather than a central determiner of the course and severity of disease. Anti-self responses have been documented in a number of models and a number of putative autoantigens have been described (Leon and Engman 2001). However there are essentially no convincing, well-controlled studies that document that anti-self responses, in the absence of infection, cause the tissue damage observed in chronic Chagas disease.

In contrast, it is very well documented that the immune responses to *T. cruzi* are highly variable in different hosts/individuals, and that the efficiency of the response dramatically effects the parasite load and the severity of disease. Thus, the explanation for the variability in chronic disease symptoms that best matches the experimental data is that individuals who make efficient, focused and potent immune responses appropriately control the infection without clinically significant destruction of host tissue. In contrast, individuals who make less effective responses have a greater chance of developing clinical disease. Those with clinical disease may have what appears to be more vigorous responses, but these responses are less efficient, less focused and yield more tissue damage while at the same time resulting in higher parasite loads in the tissues.

IMPLICATIONS AND FUTURE STUDIES

A primary implication of this model is that decreasing the stimulus for tissue damage - parasites or parasite infected cells - will lead to decreased disease severity. This conclusion is firmly supported by experiments in experimental models and in a number of long-term studies in humans. In all cases, treatments that decrease the level of parasites also significantly alter the course of disease - even when these treatments are delivered during the chronic phase of the infection (Viotti et al., 1994; Bahia-Oliveira et al., 2000; Fabbro de Suasnabar et al., 2000). Likewise treatments which fail to successfully control the level of infection also fail to modify the course of disease (Braga et al., 2000; Lauria-Pires et al., 2000).

The critical questions that this model leaves unanswered are what are the immune mechanisms which protect some individuals from disease and how can these responses be facilitated? The easy answer to the first question is that an effective anti-disease response is composed of the same set of responses that most effectively control the infection in the acute phase, i.e. antibody production and a strongly type 1 cytokine-biased response involving both CD4[+] and CD8[+] T cells. A number of studies in experimental models provide evidence that these responses are protective from disease while type 2 cytokine responses lead to more severe disease (Hoft et al., 2000; Tarleton et al., 2000; Kumar and Tarleton 2001). However much work needs to be done to determine if the same is true for human infection.

How to facilitate disease-protective responses is less clear. New and better chemotherapeutics are needed to effectively lower the parasite load in acute and chronically infected individuals. Even the rather poorly effective treatments available today show some efficacy in the treatment of chronic disease (Viotti et al., 1994; Bahia-Oliveira et al., 2000; Fabbro de Suasnabar et al., 2000). In addition, efforts should be made to develop ways to "reprogram" the immune response in individuals who are heading toward the development of severe disease because of a poorly focused or inappropriately developed anti-*T. cruzi* response. Identifying these individuals and delivering to them the best possible treatment at hand today may be the best approach for substantially altering morbidity and mortality from *T. cruzi* infection.

114 *Tarleton*

References

Almeida DR, Carvalho AC, Branco JN, Pereira AP, Correa L, Vianna PV, Buffolo E, Martinez EE. 1996. Chagas' disease reactivation after heart transplantation: efficacy of allopurinol treatment. J Heart Lung Transplant 15: 988-92.

Andrade LO, Machado CR, Chiari E, Pena SD, Macedo AM. 1999. Differential tissue distribution of diverse clones of *Trypanosoma cruzi* in infected mice. Mol Biochem Parasitol 100: 163-72.

Anez N, Carrasco H, Parada H, Crisante G, Rojas A, Fuenmayor C, Gonzalez N, Percoco G, Borges R, Guevara P, Ramirez JL. 1999. Myocardial parasite persistence in chronic chagasic patients. Am J Trop Med Hyg 60: 726-32.

Bahia-Oliveira LM, Gomes JA, Cancado JR, Ferrari TC, Lemos EM, Luz ZM, Moreira MC, Gazzinelli G, Correa-Oliveira R. 2000. Immunological and clinical evaluation of chagasic patients subjected to chemotherapy during the acute phase of *Trypanosoma cruzi* infection 14-30 years ago. J Infect Dis 182: 634-8.

Barbosa de Oliveira LC, Curotto de Lafaille MA, Collet de Araujo Lima GM, de Almeida Abrahamsohn I. 1996. Antigen-specific Il-4- and IL-10-secreting CD4+ lymphocytes increase *in vivo* susceptibility to *Trypanosoma cruzi* infection. Cell Immunol 170: 41-53.

Braga MS, Lauria-Pires L, Arganaraz ER, Nascimento RJ, Teixeira AR 2000. Persistent infections in chronic Chagas' disease patients treated with anti-*Trypanosoma cruzi* nitroderivatives. Rev Inst Med Trop Sao Paulo 42: 157-61.

Camargos ER, Franco DJ, Garcia CM, Dutra AP, Teixeira AL Jr, Chiari E, Concei, Machado CR. 2000. Infection with different *Trypanosoma cruzi* populations in rats: myocarditis, cardiac sympathetic denervation, and involvement of digestive organs. Am J Trop Med Hyg 62: 604-12.

Chen L, Watanabe T, Watanabe H, Sendo F. 2001. Neutrophil depletion exacerbates experimental Chagas' disease in BALB/c, but protects C57BL/6 mice through modulating the Th1/Th2 dichotomy in different directions. Eur J Immunol 31: 265-75.

Fabbro de Suasnabar D, Arias E, Streiger M, Piacenza M, Ingaramo M, Del Barco M, Amicone N. 2000. Evolutive behavior towards cardiomyopathy of treated nifurtimox or benznidazole and untreated chronic chagasic patients. Rev Inst Med Trop Sao Paulo 42: 99-109.

Hoft DF, Schnapp AR, Eickhoff CS, Roodman ST. 2000. Involvement of CD4 + Th1 cells in systemic immunity protective against primary and secondary challenges with *Trypanosoma cruzi*. Infect Immun 68: 197-204.

Hunter CA, Ellis-Neyes LA, Slifer T, Kanaly S, Grunig G, Fort M, Rennick D, Araujo FG. 1997. IL-10 is required to prevent immune hyperactivity during infection with *Trypanosoma cruzi*. J Immunol 158: 3311-6.

Jardim E, Takayanagui OM. 1994. Chagasic meningoencephalitis with detection of *Trypanosoma cruzi* in the cerebrospinal fluid of an immunodepressed patient. J Trop Med Hyg 97: 367-70.

Jones EM, Colley DG, Tostes S, Lopes ER, Vnencak-Jones CL, McCurley TL. 1992. A *Trypanosoma cruzi* DNA sequence amplified from inflammatory lesions in human chagasic cardiomyopathy. Trans Assoc Am Physicians 105: 182-9.

Jones EM, Colley DG, Tostes S, Lopes ER, Vnencak-Jones CL, McCurley TL. 1993. Amplification of a *Trypanosoma cruzi* DNA sequence from inflammatory lesions in human chagasic cardiomyopathy. Am J Trop Med Hyg 48: 348-57.

Kalil J, Cunha-Neto E. 1996. Autoimmunity in Chagas' disease cardiomyopathy: fulfilling the criteria at last? Parasitol Today 12: 396-399.

Kumar S, Tarleton RL. 1998. The relative contribution of antibody production and CD8+ T cell function to immune control of *Trypanosoma cruzi*. Parasite Immunity 20: 207-216.

Kumar S, Tarleton RL. 2001. Antigen-specific Th1 but not Th2 cells provide protection from lethal *Trypanosoma cruzi* infection in mice. J Immunol 166: 4596-603.

Laguens RP, Cossio PM, Diez C, Segal A, Vasquez C, Kreutzer E, Khoury E, Arana RM. 1975. Immunopathologic and morphologic studies of skeletal muscle in Chagas' disease. Am J Pathol 80: 153-62.

Lauria-Pires L, Braga MS, Vexenat AC, Nitz N, Simoes-Barbosa A, Tinoco DL, Teixeira AR. 2000. Progressive chronic Chagas heart disease ten years after treatment with anti-*Trypanosoma cruzi* nitroderivatives. Am J Trop Med Hyg 63: 111-8.

Leon JS, Engman DM. 2001. Autoimmunity in Chagas heart disease. Int J Parasitol 31: 554-60.

Marinho CR, D'Imperio Lima MR, Grisotto MG, Alvarez JM. 1999. Influence of acute-phase parasite load on pathology, parasitism, and activation of the immune system at the late chronic phase of Chagas' disease. Infect Immun 67: 308-18.

Monteon VM, Furuzawa-Carballeda J, Alejandre-Aguilar R, Aranda-Fraustro A, Rosales-Encina JL, Reyes PA. 1996. American trypanosomosis: *in situ* and generalized features of parasitism and inflammation kinetics in a murine model. Exp Parasitol 83: 267-74.

Palomino SA, Aiello VD, Higuchi ML. 2000. Systematic mapping of hearts from chronic chagasic patients: the association between the occurrence of histopathological lesions and *Trypanosoma cruzi* antigens. Ann Trop Med Parasitol 94: 571-9.

Postan M, Dvorak JA, McDaniel JP. 1983. Studies of *Trypanosoma cruzi* clones in inbred mice I. A comparison of the course of infection of C3H/HEN- mice with two clones isolated from a common source. Am J Trop Med Hyg 32: 497-506.

Rose NR. 1998. The role of infection in the pathogenesis of autoimmune disease. Semin Immunol 10: 5-13.

Salomone OA, Juri D, Omelianiuk MO, Sembaj A, Aguerri AM, Carriazo C, Barral JM, Madoery R. 2000. Prevalence of circulating *Trypanosoma cruzi* detected by polymerase chain reaction in patients with Chagas' cardiomyopathy. Am J Cardiol 85: 1274-6.

Sanz OP, Ratusnu AF, Aristimuno GG, O'Neill EM, Sica RE. 1978. An electrophysiological investigation of skeletal muscle in human chronic Chagas' disease. Arq Neuropsiquiatr 36: 319-26.

Sartori AM, Lopes MH, Caramelli B, Duarte MI, Pinto PL, Neto V, Amato Shikanai-Yasuda M. 1995. Simultaneous occurrence of acute myocarditis and reactivated Chagas' disease in a patient with AIDS. Clin Infect Dis 21: 1297-1299.

Lages-Silva E, Crema E, Ramirez LE, Macedo AM, Pena SD, Chiari E. 2001. The relationship between *Trypanosoma cruzi* and human chagasic megaesophagus. Am J Trop Med Hyg 65: 435-441.

Tarleton RL. 1997. Immunity to *Trypanosoma cruzi*. In: Host Response to Intracellular Pathogens. Kaufmann SHE (ed). Austin, TX, R.G. Landes Company: 227-247.

Tarleton RL, Grusby MJ, Postan M, Glimcher LH. 1996. *Trypanosoma cruzi* infection in MHC-deficient mice: further evidence for the role of both class I- and class II-restricted T cells in immune resistance and disease. Int Immunol 8: 13-22.

Tarleton RL, Grusby MJ, Zhang L. 2000. Increased susceptibility of *Stat4*-deficient and enhanced resistance in *Stat6*-deficient mice to infection with *Trypanosoma cruzi*. J Immunol 165: 1520-5.

Tarleton RL, Sun J, Zhang L, Postan M. 1994. Depletion of T-cell subpopulations results in exacerbation of myocarditis and parasitism in experimental Chagas' disease. Infect Immun 62: 1820-9.

Tarleton RL, Zhang L. 1999. Chagas Disease Etiology: Autoimmunity or parasite persistence? Parasitol Today 15: 94-99.

Vago AR, Macedo AM, Adad SJ, Reis DD, Correa-Oliveira R. 1996a. PCR detection of *Trypanosoma cruzi* DNA in oesophageal tissues of patients with chronic digestive Chagas' disease. Lancet 348: 891-2.

Vago AR, Macedo AM, Oliveira RP, Andrade LO, Chiari E, Galvao LM, Reis D, Pereira ME, Simpson AJ, Tostes S, Pena SD. 1996b. Kinetoplast DNA signatures of *Trypanosoma cruzi* strains obtained directly from infected tissues. Am J Pathol. 149: 2153-9.

Vago AR, Andrade LO, Leite AA, d'Avila Reis D, Macedo AM, Adad SJ, Tostes S Jr, Moreira MC, Filho GB, Pena SD. 2000. Genetic characterization of *Trypanosoma cruzi* directly from tissues of patients with chronic Chagas disease: differential distribution of genetic types into diverse organs. Am J Pathol 156: 1805-9.Viotti R, Vigliano C, Armenti H, Segura E. 1994. Treatment of chronic Chagas' disease with benznidazole: clinical and serologic evolution of patients with long-term follow-up. Am Heart J 127: 151-62.

Zhang L, Tarleton RL. 1999. Parasite persistence correlates with disease severity and localization in chronic Chagas' disease. J Infect Dis 180: 480-6.

SPECIFIC TREATMENT FOR *TRYPANOSOMA CRUZI* INFECTION (CHAGAS DISEASE)

A. Raasi* and A. O. Luquetti**
*Universidade Federal de Goiás, Goiânia
**Department of Parasitology, Instituto de Patologia Tropical e Saúde Pública, Universidade Federal de Goiás

ABSTRACT

The agent of Chagas disease is the protozoan parasite *Trypanosoma cruzi,* which causes a life-long infection in humans. The extent of damage varies according to parasite strains and individual host and may cause incapacity and even death. Nowadays, it has become predominantly held that the outcome of the disease is largely dependent upon the presence of the parasite, so that treatment with trypanocidal drugs is desirable. The elimination of this protozoan is possible with at least two available drugs, benznidazole (Bz) and nifurtimox (Nf), which have proven to be partially effective, cause acceptable side effects and are cost effective. Nevertheless, these drugs are not effective in all infected individuals and are not devoid of side effects, so treatment with these (high complexity) drugs, should be performed by physicians with specialist knowledge of the disease and of the drugs involved.

SPECIAL FEATURES OF *T. CRUZI* INFECTION

As with other infectious diseases caused by eukaryotic organisms, which are similar to human cells, safe etiological treatment is not straightforward. Few drugs are available, and success is not always guaranteed. Physicians treating a *T. cruzi* infected individual should bear in mind the following characteristics of Chagas disease.

a) The infection has two definite phases: acute and chronic. The concept of recent and late chronic infection has been introduced recently, mainly because of the good and quick response of those recently infected (less than ten years) to etiological treatment. Patients during the acute phase (up to 60 days after first symptoms) should always be treated. It is recommended to treat all those recently infected. Indications for treatment in the late chronic phase will depend on several factors analyzed below.

b) Once established, the infection lasts for life, independent of organ damage. The infection has two characteristics: the presence of the parasite and a strong humoral response against it, mainly of IgG1. Both characteristics have a different meaning and expression. Circulating parasites are the hallmark of the acute phase, but are scarce in the chronic phase. Antibodies, of low affinity during the acute phase, are not useful for diagnosis, but as the chronic phase sets in, antibodies became the hallmark of this phase, because they are nearly always present, in high titers. So, the constant feature of the chronic phase is the presence of antibodies against *T. cruzi*. If an infected individual is successfully treated during the acute or the chronic phase, parasites should disappear as well as those antibodies that were formerly present. This concept is extremely important in order to understand the basis of the follow up of etiologically treated patients. Most of all the discrepancies found in the literature on the subject depend on the correct handling of these concepts. It is obvious that, if antibodies are always present, we should look for their progressive disappearance in order to certify cure. The problem is that the parasite is extremely antigenic and so antibody responses are very strong. Different kinds of classes and subclasses are directed to components of the very heterogeneous antigenic make up of *T. cruzi*, in a relationship formed over several years to decades. As a consequence, the progressive decrease in titers usually takes years or decades until serology becomes negative , the only accepted cure criterion. On the other hand, as parasitemia is low in the chronic phase, it is quite hard to find parasites, even in those not cured, because of the low sensitivity of the methods used. Recent evidence points out that even PCR may be negative for long periods in those therapeutic failures, because the blood of infected patients may have no circulating parasites for long periods of time.

c) This protozoan has several evolutive forms, two of them in humans: the circulating trypomastigote and the intracellular amastigote, which divides by binary asexual reproduction. Any nucleated cell may be parasitized by amastigotes. The cell cycle is around 14 h (doubling time), but differences between strains have been found. After 5 to 9 divisions (4 to 7 days), new trypomastigotes (up to 512) disrupt the cell wall of the parasitized cell and circulate or penetrate another nucleated cell. Although drugs are active against both evolutive forms it is probably due to this long intracellular cycle that it is necessary to take the drug for 60 days. Experimental data showed years ago (Brener, 1961) that only mice treated for long periods of time (60 days) had the chance of cure, and those treated for a few days remained uncured.

d) Both available drugs came into use relatively recently (1965 for Nf and 1971 for Bz). So it is only a few years ago, when the first chronic phase patients reached 20 - 30 years of follow up, that it was possible to show some of them were cured. It is expected that the figures for those cured will increase as more time elapses between starting treatment and follow up performed decades later.

e) Once in the chronic phase, a proportion of those infected will have some organs affected, but nearly half will never show symptoms or organ damage. These are described as the indeterminate clinical form, also known as the asymptomatic form, characterized by normal electrocardiogram and chest, esophagus and colon X-ray. It has been estimated that around 2-5% of such cases will convert to the symptomatic form per year of infection (cardiac, megaesophagus, megacolon or associated cardiac and megaviscera).

f) Diagnosis of *T. cruzi* infection by serological methods is not 100% specific, because of cross-reactions (mainly with leishmaniasis). The physician that intends to treat a patient, should be confident of the correct etiological diagnosis, in order to avoid treatment of uninfected individuals. In order to be sure of the chagasic etiology, two serological tests of different principles are requested and the titer of antibodies evaluated. Alternatively, a positive parasitological test is accepted as an unequivocal evidence of *T. cruzi* infection.

DRUGS CURRENTLY AVAILABLE

Only two drugs are available up to now: nifurtimox and benznidazole. Both have been in use for the last four decades and considerable expertise in Latin America, has been gained with them. Several others have been tested but these two remained useful because both were effective and tolerated. This does not mean that they are effective in all cases or that they are devoid of side effects, so new drugs are hoped for. Reports on the effect of allopurinol were published (Gallerano, 1990), though further studies demonstrated that the effect in chronic phase patients was similar to placebo.

Nifurtimox (Nf)

Nifurtimox is a nitrofuran, synthesised by Bayer (Bayer 2502), introduced in 1965 with the commercial name Lampit ®. It is a [3-methyl-4-(5'-nitrofurfurylidene-amino) tetra hydro 4H-1, 4-thiazine-1, 1-dioxide]. This drug is presented in tablets with 30 and 120 mg of active substance. It is used *per os* at the dose of 8-10 mg/kg body weight/day in adults, and 10-15 mg/kg body weight/day in children every 8 hours for 60-90 days (Rassi and Luquetti, 1992; Pan American Health Organization, 1999).
The production of this drug has been discontinued since 1998, but recent information indicates that it will be produced again, now in a Latin-American country.

Side effects include anorexia with weight loss, nausea, vomiting, peripheral sensitive polyneuropathy (mainly of distal parts of the lower limbs), allergic dermopathy and several central nervous system alterations such as excitation, insomnia and seizures (rarely). Onicholysis may occur. All symptoms are reversible after drug withdrawal.

Nifurtimox was used in the acute phase and in the chronic phase, mainly in Chile, Argentina and Brazil. In this last country, results obtained were less satisfactory than in the others.

Benznidazole (Bz)

Benznidazole started to be used in 1971, and a larger experience with it has been achieved. It was synthesized by F. Hoffman-La Roche, and nominated Ro 7-1051. The commercial name is Rochagan ® in Brazil and Radanil ® in Argentina and other Spanish speaking countries. Chemically it is a N-benzyl-2-nitro-1-imidazol acetamide, with 100 mg of active substance, used orally twice a day at the dose of 5 mg/kg body weight/day for adults, and 5-10 mg/kg body weight/day in children during 60 days. Current price (Brazil) for an adult is around 15 US$ per sixty days of treatment. It has been extensively used in acute and chronic phase infected individuals.

Adverse effects of benznidazole include generalized or sometimes localized, allergic dermopathy (non-bullous polymorphic erythema) which is not dose-dependent, with slight to moderate intensity, and usually starts around the ninth day of treatment, or occasionally later. Treatment should be discontinued only when this manifestation is severe or associated with fever and lymph node enlargement. This manifestation is reversible and the use of antihistaminic drugs does not make it shorter. Another adverse effect that may appear towards the end of the treatment, and is dose-dependent, is a peripheral sensitive neuropathy, mainly of distal parts of lower limbs, when interruption of treatment is indicated. This side effect may take several months to subside and is not relieved by the administration of B-complex vitamins. At current doses, this side effect may occur sometimes, and is referred by patients as a burning pain in the soles.

Leukopenia with marked decrease in granulocytes has been rarely described, followed, sometimes, by acute tonsillitis (agranulocytopenic sore throat) which occurs 15 to 30 days after the beginning of treatment, and obviously is a formal indication for treatment withdrawal. In order to monitor this rare severe side effect, which occurs by the 20th day of treatment, it is necessary to follow up all patients clinically and by leukocyte counts. Usually, leukopenia subsides a few days after drug withdrawal and tonsillitis should be treated. Other side effects, also rarely observed and that spontaneously disappear, are loss of taste and onycholysis.

A high incidence of malignant lymphomas (8/21) was experimentally observed by Teixeira et al. (1990) in rabbits infected with the Ernestina strain of *T. cruzi* to which benznidazole was administered by intraperitoneal route at the dose of 8 mg/kg/day/60 days. No lymphomas were found in 22 control rabbits. Due to the importance of this finding, and to the lack of cases with lymphoma in several thousand patients treated with these drugs, other authors (reviewed in Cançado, 2000) performed similar experimental research in order to prove the findings stated above. No cases of lymphomas or malignant tumors were found among experimental animals in these subsequent studies.
Moya and Trombotto (1988) referred to a clastogenic effect from nifurtimox and benznidazole in three patients during treatment. This effect was observed as a high percentage of micronuclei and an increased frequency of expression of fragile sites. Nevertheless, in 6 patients with a follow up of 1 to 15 years after treatment, this effect was not observed.

To date, one of us (AR) has treated 27 patients with nifurtimox (between 1963 and 1972) and a further 160 patients with benznidazole (since 1973). These included children and adults of both sexes, in the acute and chronic phase, we have so far been unable to diagnose any malignant tumor during the long follow up. The same observation applies to a further group of 72 patients that received nifurtimox and 1,602 patients treated with benznidazole during the same period described above, but whose follow up was less strict.

In general, children show less side-effects than adults, even when higher doses during the same period of treatment are employed. Those treated during the acute phase, have a similar lack of side effects, even if adults. In adults, the prevalence of side effects is around 30% with Bz. Excretion of these drugs is mainly through the kidney.

INDICATIONS AND CONTRAINDICATIONS

Specific treatment with the available drugs is indicated in the following situations:

1) All acute phase patients, irrespective of the mechanism of transmission, including congenital transmission and accidental infection.

2) Following reactivation of the infection by immune suppression by AIDS, transplantation, treatment with corticoids and other drugs or irradiation.

3) In recent chronic phase infected individuals: these are defined by less than 10 years of infection, which means, in practice, all children, up to 12 years old. Results obtained in controlled studies with benznidazole and placebo were encouraging in this group but it is necessary to maintain the area free of vectors (bugs) in order to avoid a possible re-infection.

4) Any late chronic phase patient may be treated, after careful evaluation, if the patient will agree after the necessary explanations. Patients should be informed that treatment is not always effective, that cure may be certified after 15 - 20 years and that side effects may appear in 1/3 of the patients treated in this phase. The physician should be willing to follow up the treated individual for some years at least.

Treatment should be avoided in the following situations:
1) Pregnancy and lactation.
2) Where severe cardiopathy is already present.
3) In hepatic or renal failure, due to the impaired metabolism of these drugs.
4) In the presence of megaesophagus, which delays the absorption of the drugs, taken *per os*. Once the digestive transit has been re-established, the patient may be treated.
5) If there is hypersensitivity to these drugs.

FOLLOW UP

Using serial serology each 6/12 months until serology becomes negative. In those cured, this may take three to five years for those treated in the acute phase, five to ten years in the recent chronic phase and from 15 to 20 years in those treated during the late chronic phase. Absence of antibodies formerly present is not straightforward and antibody titers usually come down slowly, over several years. In order to compare titers, it has been very useful to store all sera, starting from samples obtained before treatment. It is recommended to store these sera in aliquots with 50% glycerin p.a. and perform the serological tests in parallel with all the samples of the patient, during the follow up. Comparisons of reactivity among the series of sera will facilitate the interpretation of the results. Tests that may be used include the immunoenzymatic assay (ELISA), indirect immunofluorescence (IIF) and indirect hemagglutination (IHA).

Parasitological tests may be used, if available. Recommended tests are xenodiagnosis, hemoculture and PCR with hybridization, this last under evaluation. If one of these tests is positive, and a false positive has been excluded, it means treatment failure and the patient should be treated with another drug or another scheme of longer duration if possible. False positive results include contamination of bugs with another kinetoplastid, *Blastocrithidia*, or contamination during the PCR procedure.

CURE CRITERIA

The accepted criterion for cure is the serology reverting to negative, i.e. the disappearance of anti-*T. cruzi* antibodies that were formerly present. Parasitological tests are useful only when positive, indicating a therapeutic failure. This means that sustained negative parasitological tests are not criteria for cure.

As the fall in titers occurs in all of those recently certified as cured, nowadays a persistent decrease in titers is very suggestive of cure. The time to certify cure depends on the phase in which the infected person was treated. Those treated during the acute phase, usually have a decline in antibody titers 6-12 months after treatment, and negative serology may be obtained in 3-5 years. If antibodies against *T. cruzi* rise after treatment, and are maintained after 5 years of follow up, the clinician should conclude that treatment was not effective. If parasitological tests are also used, a positive result might be expected during the first year (Table 1).

For children or adults treated during the recent chronic phase, a decrease in titers is observed during the first five years of follow up, and a decline to negative titres is obtained between 5 and 10 years after. Sustained antibody titers after 5 years of follow up, are indicative of therapeutic failure, but individual differences may be seen and in some cases antibody levels start to decline after 5 years.

When the late chronic phase is treated, a very slow decrease in antibody titers is observed in those who later cure. Decline in titers may start as long as ten years after treatment, and negative serology may be observed after 15 or more years. A doubtful serological result is frequently observed for long periods of time, the so called dissociated patients, with sustained negative parasitological tests and dubious serological results. Several tests have been used and claimed as better markers for cure. The first was the complement mediated lysis described by Krettli and co-workers, followed by several purified antigens tests such as the A & T ELISA, and later by recombinant protein assays, but results were controversial and these tests are still in the research phase. The only accepted criterion, so far, is the absence of anti-*T. cruzi* antibodies when measured by the so called conventional assays: ELISA, IIF and IHA all with crude antigens, extensively tested since 1975, available on the market as kits in every endemic country. It is very useful for the clinician to compare titers by each one of these tests, with the first samples, before treatment, for which it is desirable to have serum stored since the beginning.

RESULTS OBTAINED

Again, results differ in relation with the phase in which infected individuals were treated.

1) Acute phase, vertical transmission (congenital). The better results are obtained in this group when newborns are treated before one year old. Most authors record a 100% cure rate. If patients are treated later on, the percentages decline. (Moya et al., 1985). Side effects are seldom recorded.

2) Acute phase, vector transmission. A cure rate of 60 to 80% has been attained in most studies, irrespectively of the drug used (Nf or Bz). (See Rassi et al., 2000 for a review)

3) Acute phase, transfusional transmission. These patients usually have a severe disease for which transfusion was indicated (i.e. malignancy, hematological diseases), diagnosis of acute *T. cruzi* infection is usually late, and the survival rate is poor, mainly due to the severity of the primary disease on which the acute *T. cruzi* infection is superimposed. An early diagnosis of *T. cruzi* infection should improve the outcome.

4) Reactivation of the chronic phase in immunosuppressed patients. When suppression is strong (AIDS) a clinical picture of acute phase is seen, with easily demonstrable parasites in whole blood and they are frequently present in the central nervous system (search for parasites in spinal fluid). The indication is to treat them, but no results of long-term follow up are available up to now.

5) Transplantation. Both a chagasic donor or a chagasic recipient should be treated, but results on long-term outcome are not yet available.

6) Recent chronic phase. A cure rate of around 60% has been observed since 1987 (Ferreira, 1990). These earlier studies have been confirmed with two independent studies, sponsored by the World Health Organization, in Brazil and Argentina. Both studies included around 120 school children, in each study they were split into two groups: those receiving Bz and those on placebo. After a follow up of 3 to 5 years, a decrease in antibody titers was seen only in those that received Bz. Nevertheless, some children of this group did not change the antibody levels, suggesting that they were resistant, which was confirmed by xenodiagnosis in the study in Argentina. Both studies employed other non-conventional tests, which

showed completely negative serology in 56-62% of the Bz treated group. These results, together with those previously obtained, indicated treatment with Bz in school aged children was desirable in areas without active transmission (under vector-control), by which it will be expected to cure around two thirds of those treated.

7) Late chronic phase. As the expected follow up for this group is nearly 20 years, only recently was it possible to arrive at clear cut conclusions. Those few investigators who could perform this long follow up, recorded a cure rate between 8% (Cançado, 2000) and 26% (Rassi et al., in preparation). A study performed in Argentina (Viotti et al., 1994) with a shorter follow up (eight years) show declines in antibody titers only in those under Bz, as well as a lack of progression to cardiopathy in this group. All these data support recommendation of treatment for patients in the late chronic phase, on an individual basis (Pan American Health Organization, 1999). Some studies report results that do not match those already reported, such as one by Braga et al. (2000) with a small (17 treated) number of patients, all of them resistant to Nf (11) and Bz (6).

Tabular representation of different outcomes after specific treatment for *T. cruzi* infection

PHASE of DISEASE & HALLMARKS		FOLLOW UP first year	five years	ten years	twenty years	CONCLUSION
ACUTE PHASE (high parasitemia)	antibodies	yes (low titer)	no			CURE
	parasites	no	no			
	antibodies	yes (high titer)	yes (high titer)			RESISTANT
	parasites	yes	yes			
RECENT CHRONIC (high antibody titer, low parasitemia)	antibodies	yes	yes (low titer)	no		CURE
	parasites	no	no	no		
	antibodies	yes (high titer)	yes (high titer)	yes (high titer)		RESISTANT
	parasites	yes	yes	yes		
LATE CHRONIC (high antibody titer, low parasitemia)	antibodies	yes	yes	yes (low titer)	no	CURE
	parasites	no	no	no	no	
	antibodies	yes	yes	yes (high titer)	yes (high titer)	RESISTANT
	parasites	yes	yes	yes	yes	

WHERE WE ARE AND WHERE SHOULD WE GO

Great progress has been attained in the field of specific treatment for *T. cruzi* infected individuals, but we are still far from the ideal drug, which should be effective in all cases, devoid of side effects, with a single dose. Two drugs are effective in 26% to 100% of the cases, according to the phase in which the patients are treated as well as the duration of the follow up. The incidence of transmission has dropped only in some countries of the Southern Cone of South America, after vector and blood control, but we still have millions of chagasic patients who are already infected and require treatment. Specific treatment is the only alternative for those infected that will eliminate the infection and is cost effective. Nevertheless, the physician should have in mind the following issues:

1) To be sure of the chagasic etiology, by parasitological diagnosis or serology in the chronic phase. At least two conventional tests should be performed and cross-reactions with other diseases discarded.

2) Everyone in the acute phase is eligible for treatment. For recent chronic phase (often children up to 12 years old), the area should be under vector control. For late chronic phase, besides vector control, a careful evaluation of the clinical background and a clear knowledge of limitations in terms of frequency of side effects, prolonged follow up and cure rate, should be discussed carefully with each patient.

3) Bz is the drug of choice now, because most recent experience has been performed with it, side effects are less and it is the only one presently available. There is a direct relationship between duration of treatment and effect: the minimum length acceptable in order to get best results is 60 days. In case of resistance, Nf (once it is again available on the market after its withdrawal) is the alternative. New drugs are also being tested (see chapter by Urbina).

4) Monitoring antibody levels every 6-12 months should be performed as follow up. Parasitological tests are desirable, but available at present only in research institutions. They are useful only when a positive test is found, which indicates therapeutic failure.

The use of several drugs in combination will depend on their availability. The future depends on the availability of new drugs that should be more effective and with fewer side effects, an area that is still in progress.

References

Braga MS, Lauria-Pires L, Argañaraz ER, Nascimento RJ and Teixeira ARL. 2000. Persistent infections in chronic Chagas' disease patients treated with anti- *Trypanosoma cruzi* nitroderivatives. Rev Inst Med Trop São Paulo 42: 157-161.

Brener Z. 1961. Atividade terapêutica do 5-nitro-2-furaldeído-semicarbazona (nitrofurazona) em esquemas de duração prolongada na infecção experimental do camundongo pelo *Trypanosoma cruzi*. Rev Inst Med Trop São Paulo 3: 43-49.

Cançado JR. 2000. Tratamento etiológico da doença de Chagas pelo benznidazol. In: *Trypanosoma cruzi* e doença de Chagas. Brener Z, Andrade Z, Barral-Netto M. (ed.) Guanabara-Koogan, Rio de Janeiro. pp.389-405.

Ferreira HO. 1990. Tratamento da forma indeterminada da doença de Chagas com nifurtimox e benznidazol. Rev Soc Bras Med Trop 23: 209-211.

Gallerano RH, Marr JJ, Sosa RR. 1990. Therapeutic efficacy of allopurinol in patients with chronic Chagas' disease. Am J Trop Med Hyg 43:159-66.

Moya PR, Trombotto GT. 1988. Chagas' disease: clastogenic effect of nifurtimox and benznidazole in children. Medicina (B Aires) 48:487-91.

Moya PR, Paolasso RD, Blanco S, Lapasset M, Sanmartino C, Baso B, Moretti E, Cura D. 1985. Tratamiento de la enfermedad de Chagas con nifurtimox durante los primeros meses de vida. Medicina 45:553-558.

Pan American Health Organization/World Health Organization Mundial de la Salud. 1999. Tratamiento etiológico de la enfermedad de Chagas. Conclusiones de una consulta técnica. OPS/HCP/HCT/140/99. Washington.

Rassi, A. and Luquetti, A.O. 1992. Therapy of Chagas Disease. In: Chagas Disease (American Trypanosomiasis), Its Impact On Tranfusion And Clinical Medicine. Wendel S. Brener Z. Camargo, M.E., Rassi, A. (Ed.) São Paulo: Sociedade Brasileira de Hematologia e Hemoterapia. pp 237-247.

Rassi A, Rassi A Jr, Rassi GG. Fase aguda. 2000. In: *Trypanosoma cruzi* e doença de Chagas. Brener Z, Andrade Z, Barral-Netto, M. (ed) Guanabara-Koogan, Rio de Janeiro, pp.231-245.

Teixeira AR, Silva R, Cunha Neto E, Santana JM, Rizzo LV. 1990. Malignant, non-Hodgkin's lymphomas in *Trypanosoma cruzi*-infected rabbits treated with nitroarenes. J Comp Pathol 103:37-48.

Viotti R, Vigliano C, Armenti H, Segura E. 1994. Treatment of chronic Chagas' disease with benznidazole: clinical and serologic evolution of patients with long-term follow-up. Am Heart J 127: 151-162.

RATIONAL APPROACHES TO SPECIFIC CHEMOTHERAPY OF CHAGAS DISEASE

J. A. Urbina

Laboratorio de Química Biológica, Centro de Biofísica y Bioquímica, Instituto Venezolano de Investigaciones Científicas, Caracas, Venezuela

ABSTRACT

In this chapter we discuss the current status and new perspectives for the specific treatment of Chagas disease. One important development is the growing consensus around the concept that the eradication of *T. cruzi* from experimental animals and infected patients may be prerequisite to arrest the evolution of the disease and to avert its irreversible long-term consequences, which supports the use of specific anti-*T. cruzi* drugs in the management of this disease. Currently available chemotherapeutic approaches have serious limitations due to limited efficacy in the prevalent chronic form of the disease and frequent toxic side effects, but our growing knowledge of the physiology and biochemistry of *T. cruzi* is opening new perspectives for treatment. Among these new approaches, compounds that target ergosterol biosynthesis, parasite-specific proteases, trypanothione reductase, purine salvage and phospholipid biosynthesis are the most obvious potential candidates for a new specific treatment of human Chagas disease.

INTRODUCTION

As described in detail in previous chapters Chagas disease (American Trypanosomiasis), which afflicts 16-18 million persons in Latin America, is the largest parasitic disease burden of the continent and only third on a global scale after malaria and schistosomiasis. Three aspects are worth emphasizing to understand the significance of this disease, both as a public health and as a medical and scientific problem:

I)   The disease is a formally a zoonosis, very widespread among autochthonous mammals of the Americas; the parasite is transmitted among its hosts by hematophagous triatomine vectors. Human disease occurs when man invades natural ecotopes and maintains contact with the vector and reservoirs of the parasite. This means that American trypanosomiasis can never be eradicated, as its causative agent, *Trypanosoma cruzi*, is part of the biological landscape of the continent. Given the lack of success in the development of vaccines against this parasite up to the present, the only control measures available are interuption of vectorial and transfusional transmission and treatment of the persons already infected.

II)   Most of the people with *T. cruzi* infections are currently in the chronic phase of the disease and 30-40% will develop serious heart or gastro-intestinal tract lesions, due to sustained inflammatory processes triggered by the parasite. Despite a prolonged controversy associated with the possible auto-immune origin of Chagas disease pathogenesis; a growing consensus is emerging, that the eradication of *T. cruzi* from experimental animals and infected patients may be a prerequisite to arrest the evolution of the disease and to avert its irreversible long-term consequences. (See chapters by Leon and Engman and by Tarleton).

III)   Currently available chemotherapy, based on nitrofurans and nitroimidazoles (developed empirically over three decades ago) has significant activity only in the acute and short-term (up to a few years) chronic disease (see chapter by Rassi and Luquetti for details). However, as indicated above, most Chagas disease patients are in the chronic stage of the disease, where such treatments usually have low or insignificant antiparasitic action. Nevertheless, studies have shown that chronic patients subjected to antiparasitic treatment with benznidazole, although not parasitologically cured, have a detectable reduction in the occurrence of electrocardiographic changes and a lower frequency of deterioration of their clinical condition. This is probably a result of a drug-induced reduction of the parasite load in infected organs. On the other hand, nitrofurans and nitroimidazoles have significant toxic side effects, which can lead in some cases to interruption of treatment.

        Taken together, these facts indicate the urgent need of more efficacious and safer anti-*T. cruzi* treatments for the management of this important public health problem. In the remainder of this chapter I will briefly summarize the results of several research and development efforts which are being advanced towards this goal.

TOWARDS A RATIONAL APPROACH TO SPECIFIC CHEMOTHERAPY FOR CHAGAS DISEASE

Where do we stand now?

        In contrast to the empirical development of the drugs available today, efforts over the last few decades have been devoted to the study of the physiology and biochemistry of *T. cruzi*, with the aim of targeting parasite-specific biochemical pathways essential for the survival of this organism in its vertebrate hosts. Although in different stages of development, several of these approaches have shown very promising results and representative examples will be presented in the following sections.

Sterol biosynthesis and function

        T. cruzi requires, as do many fungi and yeasts, specific sterols for cell viability and proliferation and cannot use the abundant supply of the host's sterol, cholesterol. As a consequence, this organism is extremely susceptible to ergosterol biosynthesis inhibitors (EBI) *in vitro* (see Figure 1). Unfortunately, commercially available EBI, which are highly successful in the treatment of fungal diseases, such as ketoconazole, itraconazole or fluconazole (inhibitors of parasite's cytochrome P-450-dependent sterol C14a demethylase) or terbinafine (an inhibitor of squalene epoxidase) are not powerful enough to eradicate *T. cruzi* from experimentally infected animals or human patients. Thus, a breakthrough in this field was the recent demonstration that new triazole derivatives (Figure 2), such as D0870 (Zeneca Pharmaceuticals) and posaconazole (Schering-Plough), are capable of inducing parasitological cure in murine models of both acute and chronic Chagas disease, the first compounds ever to display such activity. It has further been shown that these compounds are active against both Type I and Type II *T. cruzi* and can eradicate nitrofuran- and nitroimidazole-resistant *T. cruzi* strains from infected mice, even if the hosts are immunosuppressed. It was found that these remarkable *in vivo* antiparasitic activities result from a combination of potent and selective, intrinsic anti-*T. cruzi* activity with special pharmacokinetic properties (long terminal half-life and large volumes of distribution). Moreover, these compounds are very specific for fungal or protozoan pathogens, with very low or nil toxicity towards experimental animals or humans.

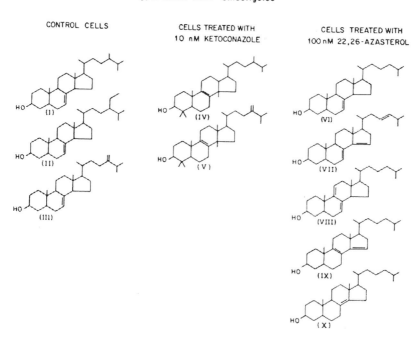

Figure 1. Endogenous sterols present in *Trypanosoma cruzi* intracellular amastigotes grown in the absence or presence of sterol biosynthesis inhibitors. Replacement of the native sterols (left panel) by its precursors (central and right panel) leads to complete growth arrest and loss of cell viability, despite the fact that these cells contain exogenous cholesterol (up to 85% by weight of their total sterol content). Ketoconazole is an inhibitor of the parasite's cytochrome P-450-dependent C14a sterol demethylase, while 22,26-azasterol inhibits 24(25)-sterol methyl transferase. I, 24-methyl-7-en-cholesta-en-3_-ol (Ergosta-7-en-3b-ol); II, 24-ethyl-7-en-cholesta-3_-ol; III, Ergosta-7,24(241)-dien-3_-ol (episterol); IV, 24-methyl-dihydro-lanosterol; V, 24-methylene-dihydro-lanosterol; VI, Cholesta-7-en-3_-ol; VII, Cholesta-7,14,24-trien-3_-ol; VIII, Cholesta-7,9(11)-en-3_-ol; IX, Cholesta-8,14-dien-3_-ol; X, Cholesta-8(14)- 3_-ol. (Modified from Urbina, 2000.).

Although development of D0870 has been discontinued, posaconazole, which is currently in Phase III clinical trials as a systemic antifungal, is a logical candidate for clinical tests with Chagas disease patients. Moreover, it has recently been found that other investigational triazoles have similar antiparasitic properties. These include (Figure 2) ravuconazole (BMS, 207,147; Bristol-Myers Squibb), TAK-187 (Takeda Chemical Company) and UR-9825 (J. Uriach and Co.). All of them are potent anti-*T. cruzi* agents *in vitro* (minimal inhibitory concentrations against intracellular amastigotes between 1 and 10 nM) with long terminal half-life in humans and several animal models. Whenever those two criteria meet, trypanocidal action results; in both acute and chronic experimental *T. cruzi* infections. Ravuconazole is particularly interesting, as its terminal half-life in humans (ca. 120 h) is by far the longest among this class of compounds. EBI of this class are currently the best candidates for a new, rational, approach to treatment of human Chagas disease.

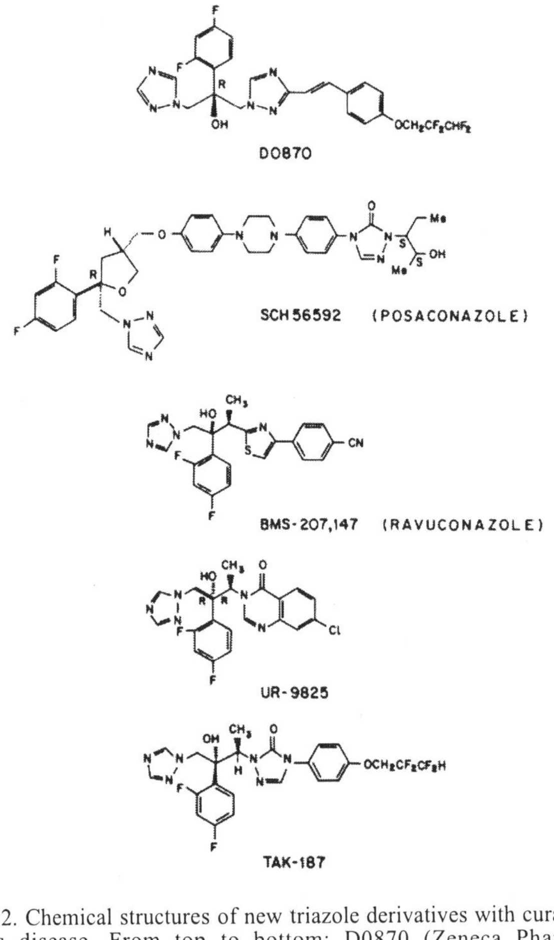

Figure 2. Chemical structures of new triazole derivatives with curative activity in experimental Chagas disease. From top to bottom: D0870 (Zeneca Pharmaceuticals), SCH 56592 (posaconazole, Schering-Plough Research Institute), BMS 207,147 (ravuconazole, Bristol-Myers Squibb), UR-9825 (Uriach and Co.) and TAK-187 (Takeda Chemical Industries). (Modified from Urbina, 2000).

Parasite-specific cysteine proteases

Cruzipain (cruzain, gp51/57) is a cathepsin L-like cysteine protease responsible for the major proteolytic activity of all stages of the life cycle of *T. cruzi*. The genes coding for this protease have been cloned and expressed, the crystal and molecular structure of the recombinant enzyme determined and structure-activity relationships of inhibitors established. Cysteine protease inhibitors (CPI), peptide analogs that inhibit specifically the activity of this protease block the proliferation of both epimastigotes and intracellular amastigotes, in the latter case with no detectable effects on host cells; such compounds also arrest metacyclogenesis (transformation of epimastigotes to metacyclic trypomastigotes) in vitro. Those multiple actions show that this enzyme performs essential functions for parasite's growth and survival. At the cellular and biochemical levels it has been shown that CPI block the maturation of cruzipain and its transport to lysosomes, as well as the breakdown of host proteins by this protease, which is essential for parasite survival. Taken together, these facts indicate that cruzipain could be an almost ideal antiparasitic target and, consistent with this hypothesis, rationally designed CPI such as N-methyl-piperazine-urea-F-hF-vinyl-sulfone-phenyl (Figure 3), are able to prolong survival and induce parasitological cure in a murine models of acute and chronic Chagas disease, with minimal toxicity. Current CPI are active both intraperitoneally and orally but have short terminal half-lives in mice, requiring relatively high doses and frequent dosing for effective activity; the search continues for compounds with improved activity and pharmacokinetic properties. Given the remarkable impact of specific protease inhibitors in the treatment of difficult infections, such as those caused by HIV, it is plausible that cruzipain-specific CPI could play in the future an important role in specific chemotherapy of Chagas disease.

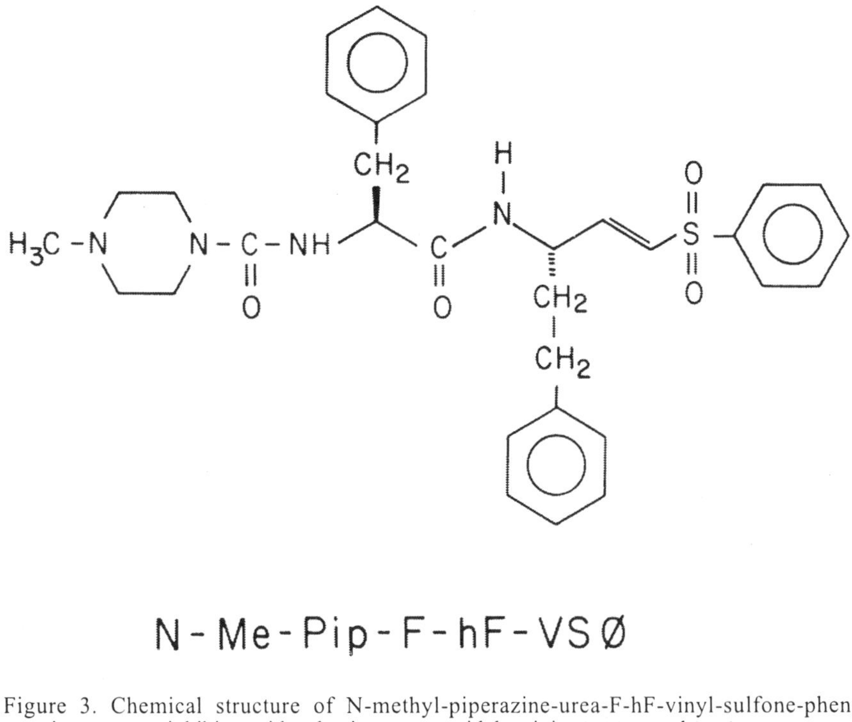

N - Me - Pip - F - hF - VS Ø

Figure 3. Chemical structure of N-methyl-piperazine-urea-F-hF-vinyl-sulfone-phenyl, a cysteine protease inhibitor with selective trypanocidal activity *in vitro* and *in vivo*.

Parasite redox metabolism

One important result of the recent interest in the study of protozoan parasites' metabolism was the discovery of trypanothione (N1,N8-bis(glutationyl)-spermidine, Figure 4) and trypanothione reductase, a unique system present in kinetoplastid protozoa, which replaces glutathione and glutathione reductase in these cells as the main intracellular thiol-redox system. Evidence that trypanothione reductase is an essential enzyme in the related organism *Leishmania donovani* has been presented; disruption of the trypanothione reductase gene, or down regulation of its expression in *L. donovani* and *L. major,* decrease their ability to survive oxidative stress in macrophages. Although, over-expression of this enzyme in *L. donovani* and *T. cruzi* does not alter their *in vitro* sensitivity to agents that induce oxidative stress such as nifurtimox, nitrofurazone and gentian violet. Thus, inhibitors of trypanothione metabolism are potential candidates for anti-*T. cruzi* drugs, alone or in conjunction with free radical-producing drugs such as nifurtimox or benznidazole.

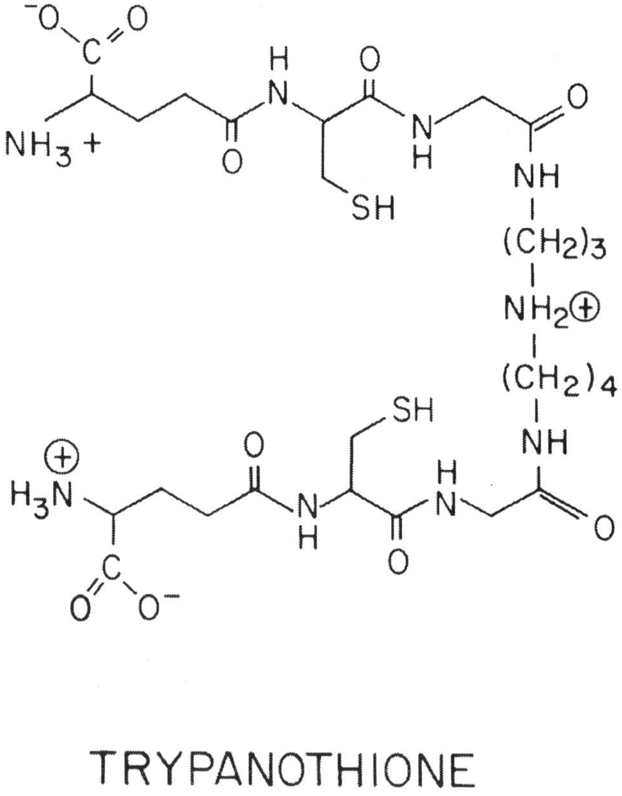

TRYPANOTHIONE

Figure 4. Chemical structure of trypanothione (N1,N8-bis(glutationyl)-spermidine), the main intracellular thiol-reducing agent in *Trypanosoma cruzi.*

Detailed structural work has been carried out with recombinant trypanothione reductase, whose molecular structure has also been solved. A series of recent studies have shown that rationally designed trypanothione reductase inhibitors display potent trypanocidal activities *in vitro*, which correlate with blockade of redox cycling in intact cells. These findings support the notion that this class of inhibitors could lead to the development of new anti-*T. cruzi* drugs.

Purine salvage pathway

Trypanosomatid parasites are absolutely deficient in the *de novo* biosynthesis of purines and dependent on scavenging of these essential compounds from the host. Hypoxanthine-guanine phosphoribosyl transferase (HGPRT) is an exclusively trypanosomatid enzyme in this pathway and is thus, in principle, a valid biochemical target in these organisms. Allopurinol (4-hydroxy-pyrazol-(3,4d)-pyrimidine) has been used for a long time in humans for the treatment of gout, as it is transformed in vertebrates into oxypurinol, a potent inhibitor of xanthine oxidase. In trypanosomatids, which are deficient in xanthine oxidase, this compound acts instead as a purine analog and is incorporated, through HGPRT, into DNA disrupting the synthesis of RNA and proteins. The compound was shown to be active against *T. cruzi* both *in vitro* and *in vivo*, but marked differences in susceptibilities among *T. cruzi* strains were also reported. There have been conflicting reports of the therapeutic efficacy of allopurinol in humans, but a multicentric study launched in 1992 in Argentina, Brazil and Bolivia was stopped as it was a patent therapeutic failure. The reasons for the lack of *in vivo* trypanocidal activity of allopurinol include low incorporation of the drug in the vertebrate stages of many *T. cruzi* strains and probably inadequate pharmacokinetic properties. The crystal structure of *T. cruzi* HGPRT is now known and recent studies, using a conformation of the enzyme approximating the transition state and a flexible docking program, were able to identify HGPRT inhibitory compounds from the Available Chemicals Directory. Several such compounds were capable of blocking the proliferation of intracellular *T. cruzi* amastigotes in cultured vertebrate cells. Such encouraging results from an *in silico* analysis suggest that in the near future HGPRT inhibitors could be useful lead compounds for the development of anti-*T. cruzi* drugs.

Phospholipid metabolism and cell signaling

Lysophospholipid analogs (LPA) are metabolically inert, synthetic analogs of lysophospholipids, which have been developed in the two last decades as anti-tumour and anti-leukaemia agents (Figure 5). These compounds have also been shown to be potent and specific antiproliferative agents against trypanosomatid parasites, including *T. cruzi*, both *in vitro* and *in vivo*. Clinical trials are underway with miltefosine, the first oral treatment for visceral leishmaniasis. Although there is significant information available on the mechanism of action of these compounds against tumoral vertebrate cells, the mechanism underlying its antiparasitic activity was not known until recently, when it was found that they selectively block phosphatidyl-choline (PC) biosynthesis in the parasite. The basis of this selective action is the fact that in *T. cruzi*, PC is synthesized thorough the Greenberg (trans-methylation) pathway, in contrast with the situation in the vertebrate host, where the Kennedy (CDP-choline) pathway is predominant; LPA are much more effective targeting the trans-methylation pathway. Thus, LPA are promising anti-trypanosomatid agents, given their potent and selective oral activity and low toxicity to hosts.

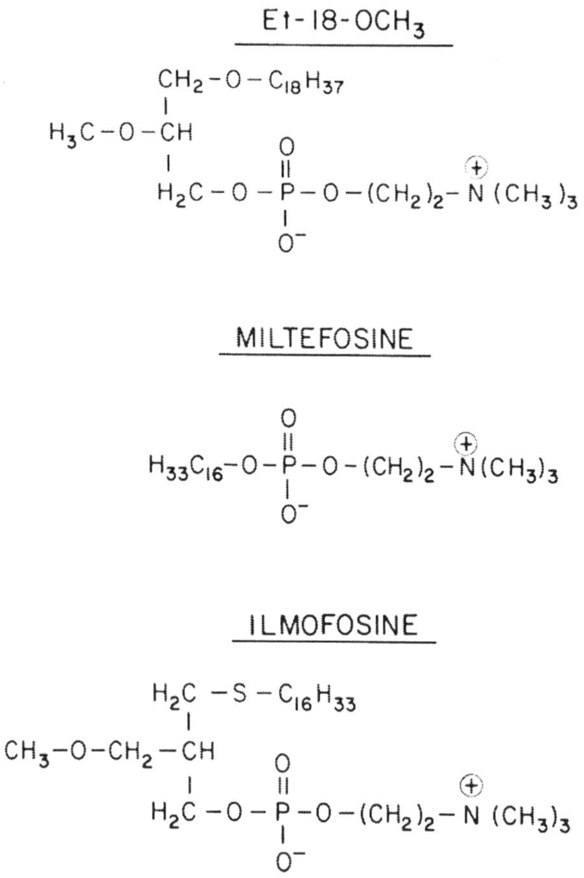

Et-18-OCH₃

$$CH_2-O-C_{18}H_{37}$$
$$|$$
$$H_3C-O-CH$$
$$|$$
$$H_2C-O-\overset{O}{\underset{O^-}{\overset{||}{P}}}-O-(CH_2)_2-\overset{+}{N}(CH_3)_3$$

MILTEFOSINE

$$H_{33}C_{16}-O-\overset{O}{\underset{O^-}{\overset{||}{P}}}-O-(CH_2)_2-\overset{+}{N}(CH_3)_3$$

ILMOFOSINE

$$H_2C-S-C_{16}H_{33}$$
$$|$$
$$CH_3-O-CH_2-CH$$
$$|$$
$$H_2C-O-\overset{O}{\underset{O^-}{\overset{||}{P}}}-O-(CH_2)_2-\overset{+}{N}(CH_3)_3$$

Figure 5. Chemical structures of lysophospholipid analogs (LPA) with anti-*Trypanosoma cruzi* activity.

CONCLUSIONS

Currently available specific Chagas disease chemotherapy, empirically developed three decades ago, is unsatisfactory due to its low efficacy in the prevalent chronic stage of the disease and frequent toxic side effects. However, a considerable amount of work has been devoted in recent decades in the identification of *T. cruzi*–specific metabolic pathways that are essential for the survival of the parasite in its mammalian hosts. This effort has begun to bear fruits with the rational development of new drugs and drug candidates with potent activity against the parasite *in vitro* and *in vivo* and very low toxicity to the host. Among these, compounds that target ergosterol biosynthesis, parasite-specific proteases, trypanothione reductase, purine salvage and phospholipid biosynthesis are the most obvious potential candidates for a new specific treatment of human Chagas disease.

FURTHER READING
Recent reviews

Croft SL, Urbina JA, Brun R. 1997. Chemotherapy of human leishmaniasis and trypanosomiasis, p. 245-257. In: Trypanomiasis and Leishmaniasis. Hide G, Mottram JC, Coombs GH, Holmes PH, CAB International, London.

Fairlamb AH, Cerami A. 1992. Metabolism and functions of trypanothione in Kinetoplastida. Ann Rev Microbiol 46:695-729.

McKerrow JH, McGrath ME, Engel JC. 1995.The cysteine proteinase of *Trypanosoma cruzi* as a model for antiparasite drug design. Parasitol Today 11:279-282.

Stoppani AOM. 1999.Quimioterapia de la enfermedad de Chagas. medicina 59 (Supl II):147-165.

Urbina JA. 1999. Chemotherapy of Chagas' disease: The how and the why. J Mol Med 77:332-338.

Urbina JA. 2000. Sterol biosynthesis inhibitors for Chagas' disease. Curr Op Anti-infect Inv Drugs 2:40-46.

World Health Organization. 1997. In: Thirteenth Programme Report, UNDP/World Bank/World Health Organization Programme for Research and Training in Tropical Diseases, World Health Organization, Geneva.

Specific articles of special interest

Bond CS, Zhang Y, Berriman M, Cunningham ML, Fairlamb AH, Hunter WN. 1999. Crystal structure of *Trypanosoma cruzi* trypanothione reductase in complex with trypanothione and the structure-based discovery of new natural product inhibitors. Structure Fold Des 7:81-89.

Croft SL, Snowdon D, Yardley V. 1996. The activities of four anticancer alkyllysophospholipids against *Leishmania donovani*, *Trypanosoma cruzi* and *Trypanosoma brucei*. J Antimicrob Chemother 38:1041-1047.

Engel JC, Doyle PS, Hsieh I, McKerrow JH. 1998. Cysteine protease inhibitors cure experimental *Trypanosoma cruzi* infection. J Exp Med 188:725-734.

Engel JC, Doyle PS, McKerrow JH. 1999.Trypanocidal effect of cysteine protease inhibitors *in vitro* and *in vivo* in experimental Chagas disease. Medicina 59 (Sup II):171-175.

Freymann DM, Wenck MA, Engel JC, Feng J, Focia PJ, Eakin AE, Craig SP. 2000. Efficient incorporation of inhibitors targeting the closed active site conformation of HPRT from *Trypanosoma cruzi*. Chem Biol 7:957-968.

Khan MO, Austin SE, Chan C, Yin H, Marks D, Vaghjiani SN, Kendrick H, Yardley V, Croft SL, Douglas KT. 2000. Use of an additional hydrophobic binding site, the Z site, in the rational design of a new class of stronger trypanothione reductase inhibitors, quaternary alkylammonium phenotiazines. J Med Chem 43:3148-3156.

Lira R, Contreras LM, Santa-Rita R, Urbina JA. 2001. Mechanism of action of antiproliferative alkyl-lysophospholipids against the protozoan parasite *Trypanosoma cruzi*. Potentiation of *in vitro* activity by the sterol biosynthesis inhibitor ketoconazole. J Antimicrob Chemother. 47:537-546.

Molina J, Martins-Filho O, Brener Z, Romanha AJ, Loebenberg D, Urbina JA. 2000. Activity of the Triazole Derivative SCH 56592 (Posaconazole) Against Drug-Resistant Strains of the Protozoan Parasite *Trypanosoma (Schizotrypanum) cruzi* in Immunocompetent and Immunosuppressed Murine Hosts. Antimicrob Agents Chemother 44:150-155.

Salmon-Chemin L, Buisine E, Yardley V, Kohler S, Debreu MA, Landry V, Sergheraert C, Croft SL, Krauth-Siegel RL, Davioud-Charvet E. 2001. 2- and 3-substituted 1,4-naphtoquinone derivatives as subversive substrates of trypanothione reductase and lipoamide dehydrogenase from *Trypanosoma cruzi*: synthesis and correlation between redox cycling activities and *in vitro* cytotoxicity. J Med Chem 44:548-565.

Santa-Rita R, Barbosa HS, Meirelles MNL, de Castro SL. 2000. Effect of alkyl-lysophospholipids on the proliferation and differentiation of *Trypanosoma cruzi*. Acta Trop 75 219-228.

Urbina JA, Payares G, Molina J, Sanoja C, Liendo A, Lazardi K, Piras MM, Piras R, Perez N, Wincker P, Ryley JF. 1996. Cure of Short- and Long-Term Experimental Chagas Disease using D0870. Science 273:969-971.

THE ECOTOPES AND EVOLUTION OF *TRYPANOSOMA CRUZI* AND TRIATOMINE BUGS

M. A Miles, M. Yeo and M. Gaunt.
Pathogen Molecular Biology and Biochemistry Unit, Department of Infectious and Tropical Diseases, London School of Hygiene and Tropical Medicine, Keppel Street, London WC1E 7HT UK.

ABSTRACT

Trypanosoma cruzi infection is a zoonosis with a complex and poorly understood epidemiology. The species *T. cruzi* is genetically diverse. Two principal subspecific groups have been identified and named *T. cruzi* I and *T. cruzi* II, the latter with five subgroups (a-e). *T. cruzi* I predominates in the Amazon basin and in endemic countries north of the Amazon; *T. cruzi* II is the predominant cause of Chagas disease throughout the Southern Cone countries of South America. We have generated *T. cruzi* I hybrids in the laboratory from clonal parental genotypes. Phylogenetic analysis based on DNA sequence data indicates that genetic exchange has contributed significantly to the evolution of *T. cruzi*. Genetic exchange may facilitate the spread of virulence and drug resistance, or the extension of host range. We propose that *T. cruzi* I may have evolved in the palm tree ecotope associated with *Didelphis* hosts and *Rhodnius* vectors, and that *T. cruzi* II may have evolved in the terrestrial ecotope with edentate hosts and *Triatoma* vectors. Comparative genomics of diverse *T. cruzi* strains may give insight into the differential pathogenesis of severe and benign Chagas disease. Molecular taxonomy and population genetics research on both *T. cruzi* and its triatomine vectors can define where domestic and sylvatic transmission cycles overlap, and contribute to the design of disease control strategies. Priorities for control of Chagas disease are improved vector control, screening of blood and organ donors, and the development of new drugs to eliminate the present burden of human infection.

TRIATOMINE VECTORS

Trypanosoma cruzi, the protozoan agent of American trypanosomiasis (Chagas disease), is thought to be the most important parasitic infection in Latin America. More than 10 million people carry the infection: drug treatment is unsatisfactory and there is no vaccine. Chagas disease is a complex zoonosis. The epidemiology is imprecisely understood.

The insect vectors of *T. cruzi* are reduviid bugs of the subfamily Triatominae (Hemiptera: Reduviidae: Triatominae). New species of Triatominae continue to be discovered and described. To date there are approximately 133 species, although a few may be synonymous. The vast majority of these species occur only in the Americas. Of the thirteen species recorded from the Old World, eight are closely related to *Triatoma rubrofasciata,* which is thought to have spread around the world with shipping and the ship rat, *Rattus rattus.* The remaining five species belong to the genus *Linschcosteus,* for which the origin is less certain, some authorities believe it to have an independent Old World origin, others consider that it may be derived from the New World tribe Triatomini. *T. rubrofasciata* carries

Trypanosoma conorhini, which is transmitted to rats via infected triatomine faeces. Rarely *T. rubrofasciata* is found infected with *T. cruzi* in Latin America but not elsewhere. It is fortunate that triatomine bugs have not spread to rural Africa, where the structure of local houses is similar to that of infested dwellings in Latin America.

The natural ecological niches (ecotopes) of triatomine bugs are wide ranging. Natural habitats include palm trees, rock piles, burrows, hollow trees, tree holes, beneath tree bark, ephiphytes and nests of some bird species. Sylvatic bugs do not fly to a moving host to take a blood meal. Rather they infest the refuges and nesting sites of their vertebrate hosts. A few species have adapted to infest houses and peridomestic dwellings. These species may form large thriving colonies, with thousands of bugs in a single house, feeding from people and their domestic animals, such as dogs, cats and guinea pigs. *Triatoma infestans* is the main vector in the Southern Cone countries of South America (Argentina, Bolivia, Brazil, Chile, Paraguay and Uruguay) and in southern Peru. *Rhodnius prolixus* and *Triatoma dimidiata* are principal vectors in northern South America and Central America. *Panstrongylus megistus* is an important vector in eastern and central Brazil, and *Triatoma brasiliensis* is domestic in northeastern Brazil (Gaunt and Miles, 2000).

VERTEBRATE HOSTS

All mammal species are thought to be susceptible to *T. cruzi* infection, whereas birds and reptiles are not. More than 150 species of mammal, comprising 24 families, are recorded as infected with *T. cruzi* or *T. cruzi*-like trypanosomes. Prevalence rates for sylvatic vertebrate hosts have seldom been determined accurately. Many host records are derived from single or small numbers of reports. Nevertheless it is clear that some mammal species, such as the marsupials *Didelphis marsupialis* and *Philander opossum*, and the edentate (armadillo) *Dasypus novemcinctus* may have high infection rates. The geographical distribution of *T. cruzi* in vertebrates and triatomine bugs is far more extensive than the range of the human disease, in part due to vector behaviour and the socioeconomic status of human populations (Miles, 1998).

A morphologically distinct trypanosome, *Trypanosoma rangeli*, is also transmitted to vertebrates in Latin America by triatomine bugs. The vectors of *T. rangeli* are bugs of the genus *Rhodnius*. In contrast to *T. cruzi*, *T. rangeli* invades the salivary glands of *Rhodnius*, so that transmission is by the bite and not by contamination with infected bug faeces. A common host of both *T. rangeli* and *T. cruzi* is the opossum, *Didelphis* spp. Molecular comparisons suggest that *T. rangeli* may be closely related to *T. cruzi*. *T. rangeli* infects humans but is considered to be non-pathogenic.

T. cruzi-like trypanosomes, of the same subgenus as *T. cruzi* (*Schizotrypanum*) are cosmopolitan in bats. Nevertheless vector borne *T. cruzi* is not known outside Latin America.

TRANSMISSION CYCLES

Triatomine species are often designated as *domestic, peridomestic,* or *sylvatic*, to indicate their biological behaviour and to represent the risk that they pose as vectors of *T. cruzi* to humans. Most triatomine species are entirely sylvatic, with little or no significance to the epidemiology of Chagas disease. Some species have quite specific habitat and host associations others are eclectic and can feed from several different host species. A few, such as

Eratyrus mucronatus, will take invertebrate blood, at least in the nymphal stages. *Panstrongylus lignarius* will descend tree trunks to approach potential hosts, presumably attracted by vibration and movement but does not fly to the host to feed. Only *Rhodnius brethesi* appears to pose a sylvatic threat, as it may contaminate forest workers gathering fibre from the piassaba palm, which is its natural habitat. Several sylvatic species may be light attracted into houses as adults. Sporadic cases of Chagas disease may result, either when an adult bug attacks inhabitants who are unlucky enough to become contaminated with infected bug faeces, or when the bugs contaminate food. Examples of species that fly to lights in the Amazon basin are *Panstrongylus geniculatus*, *Rhodnius pictipes* and *Eratyrus mucronatus*. About half of the known cases of human Chagas disease in the Amazon basin are due to small epidemic outbreaks of orally transmitted infection. Light may be used to illuminate palm or sugar cane presses at night; rarely bugs contaminate such presses and may transmit *T. cruzi* infection to those who consume the juice. Regions where *T. cruzi* is abundant in sylvatic mammals and sylvatic triatomine bugs but which have only sporadic cases of Chagas disease and no domestic bug colonies are referred to as having *enzootic transmission cycles*. Even in the Amazon basin a few triatomine species show a tendency towards domiciliation. For example *P. geniculatus* has been reported from pig sties abutting houses. Such species are sometimes considered to be *candidate domestic vectors*.

Sylvatic and domestic transmission are considered to be *continuous or overlapping transmission cycles* when the same triatomine species is found in houses and adjacent sylvatic habitats. Examples of such overlapping transmission cycles are found in northeastern Brazil, where *T. brasiliensis* infests houses and adjacent rock piles, and, probably also in some regions of Venezuela where *R. prolixus* infests houses and adjacent palm trees.

At least four of the five highly domiciliated triatomine species have spread beyond their original sylvatic range. The most notable example is *T. infestans,* which is thought to have originated from Bolivia and to have spread to all six Southern Cone countries, and to southern Peru. Similarly *R. prolixus* is thought to have spread to Central America, *T. dimidiata* from Central America to parts of northern South America and *P. megistus* from southern Brazil to central and northern Brazil. Another example of a species which has extended its range in this way is *R. ecuadoriensis* in Peru and Ecuador. Localities with a domestic triatomine species confined to houses have *discontinuous or separate transmission cycles*.

Definition of whether domestic and sylvatic transmission cycles are *overlapping* or *separate* is crucial to the planning of control campaigns. If domestic colonies are replenished from sylvatic foci spraying must be repeated more frequently and follow-up surveillance may need to be prolonged and more rigorous. If there are no relevant sylvatic populations reinvasion from sylvatic foci does not need to be considered. Molecular genetic analyses of triatomine populations and of *T. cruzi* strains can help to define whether there is continuity between domestic and sylvatic transmission cycles (Miles, 1997).

Not all *T. cruzi* transmission is vector borne. Blood transfusion transmission is a significant cause of human infection and has given rise to tens of thousands of cases of Chagas disease. Serological screening of donor blood for antibodies to *T. cruzi* is now a fundamental requirement of control programmes, and a legal requirement in many endemic countries. Similarly unscreened organ donors can transmit *T. cruzi* to recipients of organs. Immunosuppression may exacerbate the acute phase in such recipients, or may reactivate infection in recipients who were already chronic carriers to produce a secondary acute phase infection. Immunosuppression associated with HIV and AIDS may also reactivate chronic infection with severe sequelae such as meningoencephalitis, which carries a poor prognosis. Congenital *T. cruzi* infection may occur in a small proportion of infants born of seropositive mothers.

Either blood transfusion transmission or, more rarely, congenital transmission may occur far from endemic areas. The risk of blood transfusion transmission outside endemic areas is increasing with migration of populations from Latin America. Migrants who have been long term residents in rural endemic areas should be screened for antibodies to *T. cruzi* and excluded as blood donors if they are seropositive. Consumption of blood or raw meat from reservoir hosts may also give rise to oral infection. Anal gland secretions from the opossum *Didelphis* may also be a source of infection.

It is likely that the oral route of infection is extremely important in sylvatic cycles involving insectivorous hosts such as marsupials, rodents, edentates and primates, all of which may eat infected triatomine bugs.

ACUTE AND CHRONIC CHAGAS DISEASE

The initial *acute phase* of Chagas disease is often asymptomatic. Nevertheless up to ten percent of acute phase infections may be fatal. In the absence of successful drug treatment infection is usually retained for life, as is seropositivity. Spontaneous elimination of infection and reversion to seronegativity are rare. Following the acute phase most patients enter an *indeterminate phase* with no overt symptoms of disease, which in most cases lasts for life. However around thirty percent of infected individuals develop chronic Chagas disease with ECG abnormalities, sometimes associated with megaesophagus or megacolon. The geographical distribution of these latter signs of chronic Chagas disease is striking and enigmatic. It appears that megaesophagus and megacolon are very well known as a consequence of *T. cruzi* infection in central and eastern Brazil but virtually unknown in northern South America and Central America. This is one of several features of the biology and epidemiology of the *T. cruzi* zoonosis which has lead to the suggestion that *T. cruzi* may be heterogeneous with benign and virulent genotypes (Miles, 1998).

T. CRUZI GENETIC DIVERSITY

The concept of heterogeneity of *T. cruzi* arose not only from the differential geographical distribution of chronic disease but from varied success with chemotherapy, from antigenic comparisons, and distinct behaviours of *T. cruzi* strains in mice and triatomine bugs. The advent of multilocus enzyme electrophoresis (MLEE) enabled a systematic comparison of *T. cruzi* phenotypes, and genotypes (by interpretation of isoenzyme profiles). In 1977 an analysis of domestic and field isolates from Sao Felipe, Bahia State Brazil revealed distinct *T. cruzi* strains in domestic and sylvatic transmission cycles (Miles et al., 1977). This accorded with the discovery of *Triatoma tibiamaculata* in opossum refuges in bromeliad epiphytes. Not only were the domestic and sylvatic strains of *T. cruzi* distinct but they were distinct by eleven out of eighteen enzyme profiles, that is more distinct than biologically and clinically separate species of *Leishmania*. The two *T. cruzi* strains were designated zymodeme 1 (Z1) and zymodeme 2 (Z2). Subsequent studies demonstrated that *T. cruzi* Z1 was not confined to domestic cycles elsewhere, in fact it was the cause of sporadic acute cases of Chagas disease in Amazonian Brazil, and predominated in domestic transmissions cycles in Venezuela (Miles et al., 1978; Miles et al., 1981). Wider studies demonstrated that *T. cruzi* Z2 was common in domestic transmission cycles in the Southern Cone countries of South America, whereas *T. cruzi* Z1 predominated in endemic countries north of the Amazon. This led to the simplistic but unproven hypothesis that *T. cruzi* Z2 was the virulent agent of chronic Chagas disease and that *T. cruzi* Z1 was more benign, although both zymodemes were associated with similar acute phase infections and with chagasic cardiomypathy (Luquetti et al., 1986). Furthermore, the isoenzyme phenotypes reinforced that idea that domestic and sylvatic transmission cycles were largely separate in the Southern Cone region but might be overlapping in at least some parts of Venezuela.

MLEE is still an important method of characterising *T. cruzi* isolates but it has been supplemented with a wide range of DNA based techniques. These include: restriction fragment length polymorphism (RFLP) analysis of kinetoplast DNA (kDNA, schizodeme analysis); random amplification of polymorphic DNA (RAPD); comparison of ribosomal and mini-exon DNA sequence polymorphisms; microsatellite analysis, and RFLP analysis of two internal transcribed spacer regions (ITS1 and ITS2). All these approaches support the subdivision of *T. cruzi* into two major phylogenetic groups, which have now been named by international consensus as *T. cruzi* I and *T. cruzi* II. *T. cruzi* I corresponds with Z1 and *T. cruzi* II incorporates Z2 and four other subgroups of *T. cruzi* II, designated *T. cruzi* IIa-e (Tibayrenc et al., 1993; Fernandes et al., 1998; Oliveira et al., 1998).

Remarkably the genotypic subspecific groups of *T. cruzi* coincide with phenotypic groups originally defined by isoenzymes. A relatively consistent and coherent picture has emerged. Thus the current view of the subspecific taxonomy of *T. cruzi* is as follows (Brisse et al., 2000):

T. cruzi I, equivalent to isoenzyme phenotype Z1

T. cruzi II
T. cruzi IIa, isoenzyme phenotype Z3
T. cruzi IIb, isoenzyme phenotype Z2
T. cruzi IIc, isoenzyme phenotype Z3/Z1 ASAT (aspartate aminotransferase)
T. cruzi IId, isoenzyme phenotype Bolivian Z2
T. cruzi IIe, isoenzyme phenotype Paraguayan Z2

The position of Z3, as *T. cruzi* IIa, within *T. cruzi* II is somewhat controversial, some authors consider Z3 to be more closely related to *T. cruzi* I than *T. cruzi* II (Fernandes et al., 1999). The latter authors also separate Z3 into two subgroups, which may correspond with the groups designated *T. cruzi* IIa and IIc.

A significant weakness of this view of the subspecific taxonomy of *T. cruzi* is that it is based on isolates collected sporadically across vast geographical distances. The lack of large numbers of isolates from single localities has limited genetic analysis of *T. cruzi* population structures.

T. CRUZI: GENETIC EXCHANGE

Karyotype analysis indicates that *T. cruzi* is at least diploid, and this has generally been presumed to be the case in genetic analysis of *T. cruzi* population structures. Tests for random mating (panmixia) in *T. cruzi* populations by using the Hardy-Weinberg equilibrium test, and for departure from panmixia using linkage disequilibrium have consistently indicated that *T. cruzi* is substructured into asexual clonal populations. Nevertheless these data are not ideal because sample sizes are small, geographically dispersed and not from single transmission cycles. Furthermore, most studies are based on MLEE, which is less sensitive than nucleotide sequencing approaches. Population genetic analyses initially led to the conclusion that genetic exchange was extremely rare or entirely absent from *T. cruzi* populations.

In contrast to the population genetic studies some of the early isoenzyme profiles tantalisingly looked like typical heterozygous phenotypes. Some patterns in *T. cruzi* biological clones (populations derived from a single organism) were triple banded for glucose phosphate isomerase (GPI), a dimeric enzyme, and double banded for the monomeric enzyme phosphoglucomutase (PGM). These patterns were typical of those expected for heterozygotes and in some localities at least one corresponding homozygous pattern was seen. In particular, such heterozygous profiles were abundant among isolates of Bolivian Z2 (*T. cruzi* IId) and Paraguayan Z2 (*T. cruzi* IIe). This further suggested that some *T. cruzi* strains might be the product of genetic hybridisation (Miles, 1985). Strikingly, experimental studies by Dvorak and his collaborators had revealed wide variation in DNA content among *T. cruzi* strains, even in clones derived from a single strain *in vitro*. Dvorak suggested that hybridisation events might give rise to *T. cruzi* strains with increased DNA content (McDaniel and Dvorak, 1993).

In 1996 we described putative parental and hybrid PGM phenotypes of *T. cruzi* I from a single locality in the Amazon basin of Brazil. The frequencies of the PGM alleles appeared to be in Hardy-Weinberg equilibrium, although this was not statistically significant (Carrasco et al., 1996). We attempted to generate hybrids in the laboratory by passaging together *T. cruzi* clones with the two putative parental genotypes. Clones of the putative parents were first genetically (episomally) transformed to be drug resistant to either hygromycin or G418. After passage through the entire life cycle resultant populations were selected for double drug resistance, that is for the ability to grow in the presence of both hygromycin and G418. A small number of double drug resistant biological clones were obtained (Stothard et al., 1999). It was confirmed that these double drug resistant clones had hybrid characteristics, as follows:

*   Episomal constructs derived from both parents
*   A combination of parental PGM phenotypes
*   A combination of both parental cysteine protease (CP)
*   Sharing of RAPD bands between hybrids and clones

Machado and Ayala (2001) have recently used maximum likelihood phylogenetic analysis to re-examine the relationships between *T. cruzi* I and *T. cruzi* II, including its sub-groups. They concluded that the nuclear genomes of *T. cruzi* isolates representing groups IId and IIe where indeed hybrid, and that they might be derived from parental strains similar to *T. cruzi* IIb and IIc. These authors conclude that genetic exchange events across and within subgroups of *T. cruzi* II have significantly contributed to the genetic diversity and evolution of *T. cruzi*.

However, our experimental findings (above, and Gaunt et al., unpublished data) lead us to conclude that genetic hybridisation is still an active mechanism generating genetic diversity in *T. cruzi*.

HOST AND VECTOR ASSOCIATIONS

T. cruzi I predominates in domestic transmission cycles in all the endemic countries North of the Amazon basin. It is also the most abundant form of *T. cruzi* isolated from enzootic transmission cycles in the Amazon basin, and it is found in some sylvatic transmission cycles further South.

The common opossum, *Didelphis*, is associated with *T. cruzi* I over a vast geographical range. Although many mammal species are recorded as infected with *T. cruzi*, prevalence rates in *Didelphis* seem particularly high. Interestingly, it has been suggested anal gland infections in *Didelphis* might represent a primitive *T. cruzi* life cycle, although it is generally assumed that trypanosomes have evolved from insect kinetoplastids.

The fact that many *Rhodnius* species commonly carry *T. cruzi* I, and also transmit *T. rangeli*, which is often isolated from *Didelphis*, led us to speculate on the evolutionary history of *T. cruzi* I. We propose that *T. cruzi* I has evolved with *Didelphis* and *Rhodnius* in palm trees, which are the preferred habitat of most *Rhodnius* species, although there are exceptions such as *R. domesticus,* found in bromeliads and hollow trees and *R. paraensis*, described from an arboreal tree hole. It is not clear how far back in time this host-vector association may hold. Palms are thought to have arisen around 90 million years ago and marsupials were present in South America about 65 million years ago.

T. cruzi II is the predominant cause of Chagas disease throughout the Southern Cone countries of South America, and has presumably been disseminated by the spread of *T. infestans*. It is less easy to speculate on the evolutionary history of *T. cruzi* II. One suggestion is that *T. cruzi* II arose by transfer from marsupials into rodents and primates. However rodents and primates are thought to have arrived in South America 25 million years later than marsupials. Speculating on a more ancient evolutionary history we have proposed that *T. cruzi* II arose in edentates (armadillos) and reached rodents later by sharing of terrestrial habitats (burrows and rock piles). Interestingly at least twenty species of *Triatoma* have terrestrial rocky habitats (Gaunt and Miles, 2000).

Not surprisingly, the host and vector associations that we have suggested do not hold rigidly. The *T. cruzi* I/*Didelphis*/*Rhodnius*/palm tree, and *T. cruzi* II/edentate/*Triatoma*/terrestrial associations are temptingly simplistic. Nevertheless perceptions may change with fuller studies of the *T. cruzi* zoonosis. Data sets are inadequate at present to test fully the evolutionary significance of these associations. Furthermore the position of *T. rangeli* and bat trypanosomes is unclear.

MEDICAL AND EPIDEMIOLOGICAL IMPLICATIONS

It seems inevitable that the distinct *T. cruzi* genotypes are linked to both the pathogenesis of Chagas disease (Vago et al., 2000), and to transmission cycles involving different vertebrate hosts and vectors. The *T. cruzi* genome project is at present focused on the strain CL Brener, which has a *T. cruzi* IIe (hybrid) genotype (Andersson et al., 1998; Machado and Ayala,

2001). Comparative genomics across the two major sub-divisions of *T. cruzi* and the sub-groups of *T. cruzi* II may give insight in the differential pathogenesis of severe and benign Chagas disease. Epidemiological proof of virulent and avirulent *T. cruzi* strains is another matter and would require a large and rigorous investigation, covering confounders such as human genetic diversity, co-infections, nutritional status, and other factors. Simplification of genotyping methods will help to unravel the complexities of the *T. cruzi* zoonosis but only if they are applied as part of in-depth field studies with statistically significant data sets. Molecular taxonomy and population genetics can make a fundamental contribution to defining where sylvatic transmission cycles are relevant to designing control strategies. Such methods are applicable to both the disease agent and the triatomine vector. Genetic hybridisation of *T. cruzi* may have yielded recombinant genotypes with enhanced vigour and may continue to facilitate the spread of virulence, drug resistance, or the extension of host range. Molecular phylogenetics is likely to give further insight into evolutionary history of both *T. cruzi* and Triatominae. The latter should not however divert attention and resources from the priorities of perfecting vector control strategies and of devising new drugs to eliminate human infection.

References

Andersson B, Aslund L, Tammi M, Tran A-N, Hoheisel JD, Petersson U. 1998. Complete sequence of a 93.4 kb contig from chromosome 3 of *Trypanosoma cruzi* containing a strand-switch region. Genome Res 8: 811-815.

Brisse S, Barnabe C, Tibayrenc M. 2000. Identification of six *Trypanosoma cruzi* phylogenetic lineages by random amplified polymorphic DNA and multilocus enzyme electrophoresis. Int J Parasitol 30: 35-44.

Carrasco HJ, Frame IA, Valente SA, Miles MA. 1996. Genetic exchange as a possible source of genomic diversity in sylvatic populations of *Trypanosoma cruzi*. Am J Trop Med Hyg 54: 418-424.

Fernandes O, Sturm NR, Derre R, Campbell DA. 1998. The mini-exon gene: A genetic marker for zymodeme III of *Trypanosoma cruzi*. Mol Biochem Parasitol 95: 129-133

Fernandes O, Mangia RH, Lisboa CV, Pinho AP, Morel CM, Zingales B, Campbell DA, Jansen AM. 1999. The complexity of the sylvatic cycle of *Trypanosoma cruzi* in Rio de Janeiro state (Brazil) revealed by the non-transcribed spacer of the mini-exon gene. Parasitology 118: 161-166.

Gaunt M, Miles MA. 2000. The ecotopes and evolution of triatomine bugs (Triatominae) and their associated trypanosomes. Mem Inst Oswaldo Cruz 95: 5557-5565.

Luquetti AO, Miles MA, Rassi A, Rezende JM, De Souza AA, De Povoa MM, Rodrigues I. 1986. *Trypanosoma cruzi*: zymodemes associated with acute and chronic Chagas' disease in central Brazil. Trans R Soc Trop Med Hyg 80: 462-470.

Machado C, Ayala FJ. 2001. Nucleotide sequences provide evidence of genetic exchange among distantly related lineages of *Trypanosoma cruzi*. Proc Natl Acad Sci USA 98: 7396-7401

McDaniel JP, Dvorak JA. 1993. Identification, isolation and characterization of naturally occurring *Trypanosoma cruzi* variants. Mol Biochem Parasitol 57: 213-222.

Miles MA. 1998. New World Trypanosomiasis. In: Topley and Wilson's Microbiology and Microbial Infections, Vol 5, Cox FEG, Kreier JP, Wakelin D. (eds), Arnold, London, pp.283-302.

Miles MA, Toye PJ, Oswald SC, Godfrey DG. 1977. The identification by isoenzyme patterns of two distinct strain-groups of *Trypanosoma cruzi*, circulating independently in a rural area of Brazil. Trans R Soc Trop Med Hyg 71: 217-225.

Miles MA, Souza A, Povoa MM, Shaw JJ, Lainson R and Toye PS. 1978. Isozymic heterogeneity of *Trypanosoma cruzi* in the first autochthonous patients with Chagas disease in Amazonian Brazil. Nature 272: 819-821.

Miles MA, Cedillos RA, Povoa MM, de Souza AA, Prata A, Macedo V. 1981. Do radically dissimilar *Trypanosoma cruzi* strains (zymodemes) cause Venezuelan and Brazilian forms of Chagas' disease? Lancet 1:1338-1340.

Miles MA. 1985. Ploidy, heterozygosity and antigenic expression of South American trypanosomes. Parasitologia 27: 87-104.

Oliveira RP, Broude NE, Macedo AM, Cantor CR, Smith CL and Pena SDJ. 1998. Probing the genetic population structure of *Trypanosoma cruzi* with polymorphic microsatellites. Proc Natl Acad Sci USA. 95: 3776-3780.

Stothard JR, Frame IA and Miles MA. 1999. Genetic diversity and genetic exchange in *Trypanosoma cruzi*: dual drug-resistant "progeny" from episomal transformants. Mem Inst Oswaldo Cruz 94 (Supplement 1): 189-193.

Tibayrenc M, Neubauer K, Barnabe C, Guerrini F, Skarecky D, and Ayala FJ. 1993. Genetic characterisation of six parasitic protoza: Parity between random-primer DNA typing and multilocus enzyme electrophoresis. Proc Natl Acad Sci USA 90: 1335-1339.

Vago AR, Andrade LO, Leite AA, d'Avila Reis D, Macedo AM, Adad SJ, Tostes S Jr, Moreira MC, Filho GB, Pena SD. 2000. Genetic characterization of *Trypanosoma cruzi* directly from tissues of patients with chronic Chagas disease: differential distribution of genetic types into diverse organs. Am J Pathol 156: 1805-1809.

PARATRANSGENIC STRATEGIES FOR THE CONTROL OF CHAGAS DISEASE

E. M. Dotson and C. B. Beard
Centers for Disease Control and Prevention, NCID, DPD, 4770 Buford Highway, Atlanta, GA, U.S.A.

ABSTRACT
The use of insecticides in the elimination of Chagas disease vectors and therefore in the control of Chagas disease has limitations that have prompted the development of new control approaches. Paratransgenesis utilizes the genetic modification of symbiotic actinomycete bacteria found in the gut of triatomine bugs as a means to modify the gut environment of the bug, rendering it unfavorable to trypanosome development. Paratransgenic control, which has been developed for potential use in conjunction with insecticide programs, shows much promise: stable methods of transformation have been developed, antitrypanosomal genes have been tested, and a means of dispersal has been developed in the form the artificial feces, CRUZIGUARD, which mimics the natural coprophagic transfer of symbionts. Nevertheless, before a pilot field release, gene constructs will need to be optimized and questions concerning the health and environmental risks, as well as political issues, associated with the release of a genetically modified organism must be addressed.

INTRODUCTION
The traditional means of controlling Chagas disease by reducing or eliminating insect vector populations with pesticides have dramatically succeeded in the Southern Cone countries where no or few new cases of Chagas disease have been reported (Dias and Schofield, 1999; Kirchhoff, this volume). Nevertheless, problems that are inherent to chemical control of insects present significant obstacles to elimination of Chagas disease vector populations and thus to long-term control. Insecticide resistance, always considered a possible but not probable threat in bugs, has been detected in populations of two important vectors, *Triatoma infestans* in Brazil and *Rhodnius prolixus* in Venezuela (Vassena et al., 2000). Two of the greatest hurdles to complete control are 1) the inadequate treatment of homes, resulting in residual populations, and 2) the reinfestation of successfully treated homes by bugs from untreated peridomiciliary and/or sylvatic habitats. An illustration of how much of a threat these problems pose is demonstrated by the current distribution of *R. prolixus*, which was able to spread throughout Central America (except Costa Rica) from a single laboratory release, in less than 80 years (Dujardin et al., 1998). One must, therefore, ask the questions: "What happens if a house is not adequately treated, or an entire village, or several houses in different villages? How long will it take for treated areas to be reinfested?" Vigilant post-treatment surveillance and follow up insecticide applications will be necessary, and the cost of such programs may be prohibitive for some of the poorer countries. In addition, other issues associated with the use of insecticides, such as toxicity to the environment, to nontarget organisms, and to humans should not be ignored. It is within the

context of these problems associated with domiciliary insecticide use that we have investigated other potential means of controlling Chagas disease.

PARATRANSGENESIS

With the advent of many new techniques in the fields of molecular and cellular biology, it has become possible to transform genetically many organisms. While the genetic modification of insects can aid in the understanding of the genetic and biochemical basis of insect biology, the principal aim of research on the genetic modification of insects of medical and agricultural importance has been to introduce genes into insects for one of the following applications: 1) to render vector insects resistant to disease agents they may transmit, 2) to increase the effectiveness of beneficial predatory insects, and 3) to eliminate pest insect populations through improved sterile insect techniques. In order to achieve genetic transformation of insects, several key components need to be identified: a) the gene necessary to produce the desired effect in the population, b) the means of introducing the gene into the insect, c) a means of recognizing that the gene has been introduced and then finally, d) a means of spreading the gene in the target population. In paratrangenesis, these steps can be accomplished without the direct genetic transformation of the insect; instead symbiotic bacteria are genetically modified to achieve the desired effect. An advantage to such an approach for transformation is the ease and reproducibility of microbial transformation systems. The requirements for a paratransgenetic system to work in the control of a vector-borne disease have been described by Beard et al., 2002. These include, among other requirements, a symbiotic relationship between bacteria and insect, bacteria that are amenable to genetic manipulation, effective interaction of the transgene product with the target pathogen, and a natural dispersal method for the genetically altered symbionts. With most Chagas disease vectors, such symbiotic bacteria occur in close proximity to the parasite *Trypanosoma cruzi*. In this chapter we briefly discuss the ecology of the bacteria, the progress made in the genetic modifications of these bacteria toward their use in paratransgenic control of Chagas disease, and preliminary studies on the dispersal mechanism as well as safety concerns involved in the release of genetically modified organisms.

SYMBIOTIC RELATIONSHIPS OF TRIATOMINE BUGS AND BACTERIA

Because bugs of the family Reduviidae subfamily Triatominae feed on blood throughout their entire developmental cycle, they harbor in their gut lumen symbiotic bacteria that provide nutrients missing from their diet. Depending on the triatomine species, the gut environment may contain a pure culture of gram-positive actinomycete bacteria or a mixture of various gram-positive and -negative organisms that together provide the essential nutrients. The actinomycete symbionts of *R. prolixus* were once thought to produce pantothenic acid, a B-complex vitamin that is found in plant and animal tissues and is essential for a cellular growth (reviewed by Dasch et al., 1984). The current hypothesis is that the bacteria themselves are a part of the bug diet, with digestion of the bacteria cellular components providing the necessary nutrients (Beard et al., 2002).

Newly hatched nymphs are transiently aposymbiotic and must acquire symbionts by feeding on parental fecal material (coprophagy) found in the immediate environment. The need for these symbionts has been ascertained by rearing aposymbiotic nymphs under sterile conditions; nymphs that do not have access to symbiotic bacteria fail to molt to the adult stage. Coprophagy insures transmission of the essential symbionts from adult to progeny. Additionally, it provides an easy dispersal method for transgenic bacteria (see later in chapter).

The bacteria are fairly easy to isolate. Fecal drops can be collected from surface-sterilized bugs, suspended in culture media and plated, or bacteria can be cultured from the gut contents of bugs several days after feeding. In fifth instar nymphs of *R. prolixus* 5-7 days after the blood meal, the numbers of *Rhodococcus rhodnii* in the midgut increase to 10^9 (Dasch et al., 1984). Several species of symbiotic bacteria have been isolated from other Chagas disease vectors, including *T. infestans*, *Triatoma dimidiata, and Triatoma sordida. Rhodnius prolixus,* however, appears to have a more intimate relationship with its symbiotic bacteria *R. rhodnii;* consequently, it is with *R. prolixus* and its symbiont *R. rhodnii* that most of the work has been conducted. Therefore, this symbiotic association will be the main focus of the rest of this discussion.

TRANSFORMATION OF *R. RHODNII*

The initial transformation studies with *R. rhodnii* involved the development of a shuttle plasmid, a circular extrachromosomal element that would replicate both in *R. rhodnii* and in the laboratory workhorse *Escherichia coli* (Beard et al., 1992). The *E. coli* origin of replication and ampicillin and thiostrepton antibiotic resistance markers were derived from the plasmid pIJ30. To form the plasmid pRr1.1, the *R. rhodnii* replication origin was cloned into this plasmid from a restriction enzyme fragment digested from an uncharacterized endogenous plasmid isolated from *R. rhodnii* ATCC strain 35071. Originally protoplasts of *R. rhodnii* were used for transformation experiments, but these proved very fragile, and it was later shown that mid-log growth of bacteria could be transformed efficiently by electroporation (Beard and Aksoy, 1997). Thiostrepton efficiently kills untransformed *R. rhodnii,* and by selecting on thiostrepton agar plates after transformation, it was demonstrated that these actinomycete bacteria could be easily genetically engineered. The transformed *R. rhodnii* were fed to *R. prolixus* aposymbiotic first instar nymphs in blood through an artificial membrane. Two groups bugs were fed monthly with one group receiving thiostrepton and the other blood with no antibiotics. The survivability of the bugs was equivalent to that of bugs who had received wild-type bacteria. In addition, these genetically modified bacteria survived throughout the life of the insect with or without antibiotic selection, thus demonstrating that the genetically modified bacteria could successfully colonize *R. prolixus* and provide the normal symbiotic needs (Beard et al., 1992).

The next step was to find an antitrypanosomal gene to be expressed by symbiotic bacteria in the paratransgenic bug. The insect immune peptide cecropin A efficiently kills invading bacteria in the hemolymph of an injured saturniid moth *Hyalophora cecropia* by producing pores in the cell wall of the bacteria (Christensen et al., 1988). This peptide is also toxic to *E. coli, T. cruzi,* and log-phase *R. rhodnii* at concentrations of 23 µM, 150-240 µM, and 500 µM, respectively, but nonvegetative forms of *R. rhodnii* are resistant to any amount of cecropin (Durvasula et al., 1997). Thus, cecropin A could potentially serve as an antitrypanosomal product in a paratransgenic system. The plasmid pRr1.1 was modified to form pRrThioCec by adding the gene for the mature peptide of cecropin A. Western blotting lysates of bacteria transformed with this plasmid indicated that the bacteria produced the peptide. Aposymbiotic *R. prolixus* first instar nymphs were fed wild-type or transformed *R. rhodnii*. The growth rate and survivability to adult stage were comparable in both groups suggesting that the presence of the cecropin peptide within the gut did not affect the growth of the insects. The presence of the cecropin A peptide in the hindgut of the bugs was demonstrated in western blots and in *E. coli* killing assays of gut contents. The final test was to determine the effect of *in vivo* production of cecropin A on the numbers of metacyclic trypomastigotes in the insect hindgut. Fourth instar nymphs were fed epimastigotes in blood through an artificial membrane. The adults were dissected and in 65 % of the bugs with cecropin producing symbionts the trypanosomes were eliminated from the hindgut. In the remaining 35 %, the numbers of *T. cruzi* in the gut were reduced by 100 fold (Durvasula et al., 1997; Beard et al., 2002). Thus the constitutively expressed recombinant cecropin A from the symbionts was lytic to *T. cruzi,* providing the first demonstration of genetically altered bacteria preventing the possible transmission of a vector-borne disease.

One drawback to the use of cecropin A is that its activity is not limited to trypanosomes. In addition, trypanosomes may eventually develop resistance to cecropin expressed in the insect hindgut. Developing a larger arsenal of antiparasitic agents and developing agents that are directed more specifically to the target parasite, the trypanosome, are needed. Recently phage display technology has made it possible to clone single-chain antibody genes and express them in a heterologous system, thus allowing mammalian-derived antipathogen antibodies to be expressed in insects.

To test the feasibility of antibody secretion in the paratransgenic situation, the gene for a single chain antibody against progesterone (rDB3) was cloned into a plasmid (pRrMDWK6) so that it was under the control of a heterologous promoter/signal peptide derived from an alpha antigen gene of *Mycobacterium kansasii*. A protein comparable in size to the recombinant DB3 was detected by western blot in the supernatants of cultures of *R. rhodnii* transformed with this plasmid. Binding activity of the DB3 antibody to progesterone in the hindgut of *R. prolixus* containing the genetically modified bacteria was demonstrated by ELISA. This study was the first demonstration that a functional antibody could be secreted by a bacterium in the gut of an insect and serves as a model for antibody fragments that could be directed against key surface antigens of *T. cruzi* to inactivate the trypanosome in the gut of the bug (Durvasula et al., 1999a).

IMPROVEMENTS IN GENETIC MODIFICATION

Integrative plasmids

One of the drawbacks to using episomally located plasmids in a paratransgenic system is that in some cases the overall plasmid loss per generation from a transformed population is as much as 0.5 %. For any long-term field use, such a loss would be unacceptable and higher stability rates are necessary for long-term sustainability. Other methods of genetically modifying bacteria are being investigated, and the most promising are a specific group of integrative shuttle plasmids (plasmids that replicate episomally in *E. coli,* but integrate into the mycobacterial genome) that have been developed for use in studying mutagenesis in mycobacterial species. In these plasmids, the integrative elements (*attP* and *int* gene) of the *L5* mycobacteriophage have been cloned. In the presence of a host factor from the bacterium, the viral integrase protein cleaves the plasmid at the *attP* and the bacteria chromosome at a complimentary site (*attB*) and the plasmid is inserted into the chromosome as a viral genome would be inserted. Foreign genes such as antibiotic resistance genes and beta-galactosidase were cloned into mycobacteria in this manner and were shown to be highly stable even in the absence of antibiotic selection (Lee et al., 1991; Stover et al., 1991).

The integrative plasmid pBP5 contains the integrative elements of the *L1* mycobacteriophage as well as a hygromycin resistance gene and the *lacZ* gene. Although originally designed for use in mycobacteria, this plasmid integrates successfully into the genome of *R. rhodnii*; the transformation efficiency, however, is much lower than that of mycobacteria but is sufficient to allow construction of recombinants that stably express foreign genes. As in the mycobacteria, the plasmids were inserted only once per *R. rhodnii* genome. Nevertheless, the *lacZ* gene was expressed intensely, suggesting that the use of a strong constitutive promoter will allow ample expression of inserted genes. In addition, the plasmid was highly stable with no loss of insert over 100 generations (Dotson, E., et al., submitted). Preliminary studies of similar plasmids in several actinomycete bacteria from other triatomine bugs have demonstrated an equivalent ease of use and stability (Eichler, S., Dotson, E., Pennington, P., unpublished data). Thus, the integrative systems appear to be a superior mode of transformation for a paratransgenetic system.

Cecropin A improvements

In the plasmids containing cecropin, some degree of construct instability and possible toxicity was noted. In the originally constructed shuttle plasmid, the mature peptide was released inside the bacterium, a condition that may have been detrimental to the transformed *R. rhodnii*. Under normal conditions, a signal peptide allows transport of the molecule across the cell membrane. The signal sequence is cleaved during transport, resulting in an inactive propeptide that circulates free in the insect hemolymph. The propeptide is activated by proteolytic cleavage by a dipeptidyl-peptidase that is released as a result of immunologic stimulation. By using an appropriate bacterial promoter and signal peptide, the inactive propeptide should be released outside of the bacterium and activated against the trypanosome by the digestive enzymes in the gut of the triatomine bug. Studies are currently in progress to evaluate this proposed mechanistic improvement.

Marker genes

With the introduction of integration vectors, colorimetric markers were added to facilitate the screening for transformed colonies. The *lacZ* gene with an *hsp60* promoter from mycobacteria produces beta-galactosidase that breaks down the sugar X-gal to produce a blue color in the transformed colonies. Unfortunately, because this gene is found in many bacteria, the eventual screening of environmental samples for the transformed *R. rhodnii* may become complicated. The colorimetric marker, green fluorescent protein (GFP) has become the reporter gene of choice in transgenic studies of many insects (Higgs and Lewis, 2000). This protein was isolated from the jellyfish *Aequorea victoria,* and the GFP chromophore emits fluorescence when exposed to light of appropriate wavelengths. This fluorescence is highly sensitive and occurs without the need for exogenously added substrates or cofactors. It has the advantage of being a marker that is not likely to be found in terrestrial environments. The protein has been expressed in various mycobacteria (Via et al., 1998) and attempts are being made at expressing it in *R. rhodnii*.

Antibiotic resistance markers such as the kanamycin resistance marker, as well as the thiostrepton and hygromycin mentioned previously, have been included in the shuttle plasmids to facilitate selection of transformed bacteria. Transformed bacteria that would be considered for release, however, would not harbor antibiotic resistance genes.

STEPS TO RELEASE

Strategies for the spread of the transgenic symbionts should mimic the naturally occurring method of bacterial transfer in insects. Although in the initial experiments nymphal *R. prolixus* were fed the genetically modified bacteria through an artificial membrane, exposure of the bugs to artificial feces on filter paper was later shown to be a more efficient method of infecting the bugs with bacteria (Beard, C. B., unpublished results) and more closely reproduced the natural coprophagic route of symbiont acquisition. The artificial feces, called CRUZIGUARD, is composed of genetically modified bacteria resuspended in phosphate buffered saline, Guar gum to give the mixture substance, and India ink to approximate the dark color of feces.

Laboratory experiments to determine the efficacy of CRUZIGUARD exposure to nymphs already infected with wild-type bacteria showed that 100 % of the bugs picked up the transformed bacteria but these bacteria represented only 1 % of the total gut bacterial flora (Durvasula et al., 1999b). This low percentage was expected, because experiments of sequential exposures to wild-type and recombinant bacteria demonstrated that the initial coprophagic infections predominate throughout development (Beard, C.B, unpublished results).

In a preliminary study that would approximate the method of impregnating houses in the field, hut materials (thatch or adobe) along with soil from the field were placed in large Lucite boxes. CRUZIGUARD was applied monthly to the cracks and corners of the thatch and adobe. Females from the field were released into the boxes and were removed when their eggs hatched (Durvasula et al., 1999b). Bugs were assayed at the third and fifth nymphal and adults stages. Approximately 50 % of the bugs carried the genetically modified bacteria and these bacteria made up around 90 % of the gut flora. Thus, in this simulated field trial, transformed *R. rhodnii* could compete with other bacteria to become the established symbiont.

In an ongoing greenhouse trial to simulate a larger scale field condition, colony females have been released into a 6 ft x 6ft x 6ft hut composed of plywood and thatch. CRUZIGUARD was applied after the eggs began hatching. Preliminary evaluations indicate that at least 50 % of bugs acquire the genetically modified symbiont in this field trial design.

The establishment of a paratransgenic population of bugs that are refractory to trypanosomes will have a greater chance of success if the likelihood that aposymbiotic nymphs encounter the recombinant bacteria before they encounter native bacteria is increased; thus, the probability that the recombinant bacteria will become the predominant symbiont also increases. Several things can be done to achieve this. First, eliminate or reduce the existing population of bugs; the paratransgenic system would work best if used in conjunction with an insecticide program. Second, add to CRUZIGUARD chemo-attractants similar to compounds in the feces such as ammonia that attract nymphs to natural feces; this may increase the efficacy of CRUZIGUARD, allowing less frequent applications and lower concentrations of recombinant bacteria. Third, strategically place the CRUZIGUARD in the cracks of the adobe and thatch, the areas where eggs usually are deposited, and apply sufficient quantities of bacteria to greatly outnumber the native species; this will statistically increase the chance that the newly hatched first instar nymphs will encounter recombinant bacteria.

This approach to reducing Chagas transmission by the production of a refractory bug population is best suited for situations where vectors are widely distributed and where repeated insecticide applications may be impractical due to reinvasion. CRUZIGUARD could potentially be applied to new homes or insecticide treated homes. The progeny of the invading bugs will acquire the recombinant bacteria, thus resulting in the establishment of a population of refractory bugs. This population of bugs will spread the bacteria with antitrypanosomal genes in their fecal droplets. In this manner, paratransgenesis could be used along with insecticides, as a part of an integrated pest management program for Chagas disease control, providing a potential solution to the problems associated with reinfestation of insecticide-treated homes.

SAFETY CONCERNS

The safety concerns for the release of transgenic bacteria for control of Chagas disease are real and complex, and need to be addressed in detail before the bacteria can be released for pilot studies in the field. The risks associated with the release of genetically modified insects has been reviewed in several papers including Aultman et al., 2000; Beard et al., 2000, 2002; Hoy, 2000. Three major areas of concern are human health, environmental effects, and political and public perceptions.

The native forms of the genetically modified bacteria that could be released are normally found among the bacterial flora in the households where bugs are found. In addition, the bacteria are not known to be pathogens to mammals. These two facts suggest that the bacteria should not be a human health concern; however, because large numbers will be released, the question that has been raised is whether these bacteria could cause disease in immunocompromised or immunosuppressed individuals. In preliminary studies to begin answering this question, normal and immunocompromised nude CD1 mice were inoculated with 10^7 genetically modified *R. rhodnii* by gastric lavage. Bacteria were shed in the mouse feces for 2-4 days. Ten days after inoculation, necropsy of both groups of mice revealed no lesion or any other signs of disease in the intestinal tract or other internal organs. These data suggest that oral exposure probably will not be a health concern. More studies will be required to test for possible effects of aerosol inoculations and possible allergic reactions following chronic exposures. Initial work, however, confirms the accepted view that these bacteria are, in fact, nonpathogenic.

The environmental concerns include unanticipated effects on nontarget organisms and the risk of horizontal transfer of the introduced genes to other microorganisms resulting in unanticipated altered phenotypes. To minimize unexpected results in the field, numerous laboratory experiments are planned with nontarget organisms, including the analysis of the bacterial flora of other insects found in the homes and the inoculation of these insects with genetically modified bacteria to determine if the bacteria can colonize other insects. Experiments are planned to determine the risks of horizontal transfer of the inserted genes by culturing transformed bacteria with other species of common soil bacteria, under various conditions. Given the promiscuous nature of bacteria to exchange genetic material, horizontal transfer of some portion of the introduced genes probably will occur eventually, and even a rare event could have an ecological significance. Therefore, the focus of risk assessments should be on the transgenes themselves and the potential magnitude of the risk of the transfer of a specific sequence. To minimize risks, antibiotic resistance markers and any unnecessary gene or gene fragment should be removed. Ultimately, however, the question of releasing the transgenic bacteria will have to be weighed against the risk of acquiring disease in the current situation of inadequate control.

The regulations involved with release of genetically modified organisms have been the subject of several recent reviews (Beard et al., 2002, Hoy, 2000 and Young et al., 2000). Because Chagas disease is primarily a disease of the developing world, a pilot release would most likely occur in a third world country. Therefore, the local scientists and ministries of health of the particular country should be involved in conducting the field-stage tests. The public should be informed at the earliest time about the potential releases and should be brought into discussions about such releases.

References

Aultman KS, Walker ED, Gifford F, Severson DW, Beard CB, Scott TW. 2000. Managing risks of arthropod vector research. Science 288: 2321-2.

Beard CB, Aksoy S. 1997. Genetic manipulation of insect symbionts, In: Molecular biology of insect disease vectors, a methods manual. Crampton JM, Beard CB, Louis C. (eds.) Chapman and Hall, London, pp 555-560.

Beard CB, Mason PW, Aksoy S, Tesh RB, Richards FF. 1992. Transformation of an insect symbiont and expression of a foreign gene in the Chagas' disease vector *Rhodnius prolixus*. Am J Trop Med Hyg 46: 195-200.

Beard CB, Durvasula RV, Richards FF. 2000. Bacterial symbiont transformation in Chagas disease vectors, In: Insect Transgenesis: Methods and Applications. Handler AM, James AA. (eds.), CRC Press, New York, pp 289-303.

Beard CB, Cordon-Rosales C, Durvasula RV. 2002. Bacterial symbionts of the Triatominae and their potential use in control Chagas disease transmission. Ann Rev Entomol 47: 123-141.

Christensen B, Fink J, Merrifield RB, Mauzerall D. 1988. Channel-forming properties of cecropins and related model compounds incorporated into planar lipid membranes. Proc Natl Acad Sci USA 85:5072-6.

Dasch GA, Weiss E, Chang K. 1984. Endosymbionts of insects, In: Bergey's Manual of Systematic Bacteriology. Vol. 1., Krieg NR. (ed.) Williams & Wilkins, Baltimore, MD, pp. 811-833.

Dias J, Schofield C. 1999. The evolution of Chagas disease (American trypanosomiasis control after 90 years since Carlos Chagas'discovery. Mem Inst Oswaldo Cruz 94 (Suppl. 1):103-21.

Dujardin JP, Munoz M, Chavez T, Ponce C, Moreno J, Schofield CJ. 1998. The origin of *Rhodnius prolixus* in Central America. Med Vet Entomol 12(1): 113-115.

Durvasula RV, Gumbs A, Panackal A, Kruglov O, Aksoy S, Merrifield RB, Richards FF, Beard C.B. 1997. Prevention of insect-borne disease: an approach using transgenic symbiotic bacteria. Proc Natl Acad Sci USA 94(7): 3274-8.

Durvasula, R.V., Gumbs, A., Panackal, A., Kruglov, O., Taneja, J., Kang, A.S., Cordon-Rosales, C., Richards, F.F., Whitham, R.G., Beard, C.B. 1999a. Expression of a functional antibody fragment in the gut of *Rhodnius prolixus* via transgenic bacterial symbiont *Rhodococcus rhodnii*. Med Vet Entomol 13(2): 115-119.

Durvasula RV, Kroger A, Goodwin M, Panackal A, Kruglov O, Taneja J, Gumbs A, Richards FF, Beard CB, Cordon-Rosales C. 1999b. Strategy for introduction of foreign genes into field populations of Chagas disease vectors. Ann Entomol Soc Am. 92(6): 937-943.

Higgs S, Lewis DL. 2000. Green fluorescent protein as a marker for transgenic insects, In: Insect Transgenesis: methods and applications. Handler AM, James AA. (eds.) CRC Press, New York, pp 93-108.

Hoy MA. 2000. Deploying transgenic arthropods in pest management programs: risks and realities, In: Insect Transgenesis: Methods and Applications. Handler AM, James AA. (eds.) CRC Press, New York pp 335-367.

Lee MH, Pascopella L, Jacobs WR, Hatfull GF. 1991. Site-specific integration of mycobacteriophage L5: integration-proficient vectors for *Mycobacterium smegmatis*, *Mycobacterium tuberculosis*, and bacille Calmette-Guerin. Proc Natl Acad Sci USA 88: 3111-3115.

Stover CK, de la Cruz VF, Fuerst TR, Burlein JE, Benson LA, Bennett LT, Bansal GP, Young JF, Lee MH, Hatfull GF, Snapper SB, Barletta RG, Jacobs WR Jr, Bloom RB. 1991. New use of BCG for recombinant vaccines. Nature 351: 456-460.

Vassena CV, Picollo MI, Zerba EN. 2000. Insecticide resistance in Brazilian *Triatoma infestans* and Venezuelan *Rhodnius prolixus*. Med Vet Entomol 14: 51-55.

Via LE, Dhandayuthapani S, Deretic D, Deretic V. 1998. Green fluorescent protein, a tool for gene expression and cell biology in mycobacteria. In: Mycobacteria Protocols. Parish T, Stoker NG (eds), Humana Press, Totowa, New Jersey, pp. 245-260.

Young OP, Ingebritsen SP, Foudin AS. 2000. Regulation of transgenic arthropods and other invertebrates in the United States, In: Insect Transgenesis: Methods and Applications. Handler AM, James AA. (eds.) CRC Press, New York pp 369-379.

CURRENT PUBLIC HEALTH CONCERNS

L.V. Kirchhoff
Internal Medicine (Infectious Diseases) and Epidemiology, University of Iowa, 4-403 BSB
Iowa City, Iowa 52242, USA

ABSTRACT

American trypanosomiasis, or Chagas disease, is caused by the protozoan parasite, *Trypanosoma cruzi*. Sixteen to eighteen million people in Mexico, Central America, and South America are infected with this organism, and 45,000 deaths are attributed to the disease each year. Infection with *T. cruzi* is life-long, and 10-30 % of persons who harbor the parasite chronically develop cardiac and gastrointestinal problems associated with the parasitosis. Although major progress has been made in recent years in reducing vector-borne and transfusion-associated transmission of *T. cruzi*, the burden of disability and death in persons chronically infected with the organism continues to be enormous. Congenital Chagas disease also is a public health concern, as 5-10 % of infants born to *T. cruzi*-infected women are in turn infected, and no medications are available for preventing congenital transmission. *T. cruzi* can also be transmitted by transplantation of organs harvested from persons with Chagas disease. Eight to ten million persons born in countries in which Chagas disease is endemic currently reside in the United States, and epidemiologic and census data suggest that 50,000-100,000 are chronically infected with *T. cruzi*. The presence of these infected persons poses a risk of transmission of the parasite here through blood transfusion and organ transplantation, and seven such cases are known to have occurred.

CURRENT PUBLIC HEALTH CONCERNS

American trypanosomiasis, or Chagas disease, is caused by the protozoan parasite, *Trypanosoma cruzi*, while sleeping sickness results from infection with *Trypanosoma brucei gambiense* or *Trypanosoma brucei rhodesiense*, two subspecies of the group of protozoa called the "*Trypanosoma brucei* complex." Although these two groups of organisms belong to the same genus, much about them is different. *T. cruzi* is found only in the Americas, while the ranges of *T. b. gambiense and T. b. rhodesiense* are limited to Africa. The vectors of the two groups of organisms also are different – several genera of triatomine insects transmit *T. cruzi* and tsetse flies carry the agents of sleeping sickness. Infective forms of *T. cruzi* are found in the feces of its triatomine vectors and transmission is contaminative, while African trypanosomes are present in the saliva of the tsetse flies and transmission is inoculative. Things are also quite different in the mammalian hosts that harbor these organisms. In the many species of mammals that *T. cruzi* parasitizes, the organisms multiply intracellularly and thus avoid the onslaught of antibodies and complement. On the other hand *T. brucei* has chosen to proliferate in the bloodstream and survives immune destruction by periodically changing its coat of glycoproteins through a process called antigenic variation.

Not surprisingly, the clinical manifestations in humans of infections with these two groups of parasites are largely distinct, both in the acute and chronic forms of the illnesses. Congenital and transfusion-associated transmission of *T. cruzi* are common but occur extremely rarely with African trypanosomes. Most persons with chronic *T. cruzi* infections never develop symptoms related to the parasitosis, and the 10-30 % of infected persons who become symptomatic suffer from cardiac and gastrointestinal problems which are largely managable by medical intervention. In contrast, patients with *T. b. gambiense* and *T. b. rhodesiense* infections who do not receive specific treatment develop chronic neurologic impairment that invariably results in death. Different drugs are used to treat the two forms of trypanosomiasis, and the only commonality of the various agents is their ability to cause serious side effects. The management of the cardiac and gastrointestinal manifestations of chronic Chagas disease is supportive, and elimination of the parasites is difficult and of questionable usefulness. In patients with sleeping sickness the goal of treatment is to eliminate the infecting organisms completely and resolution of clinical problems usually follows.

Lastly, the current public health concerns relating to African and American trypanosomiases are vastly different. Althouth sleeping sickness is much less of a public health problem now than it was in the 19th and early 20th centuries when it claimed hundreds of thousands of lives, it still is a major problem. Moreover, in recent years the number of new cases has increased enormously as major epidemics have occurred in several countries, including, Cote d'Ivoire, Angola, the Sudan, and the Central African Republic, among others. Optimal approaches for reducing the problem of sleeping sickness through tsetse control of have been difficult to define and, despite decades of efforts in many regions, effective control has remained an elusive goal (See Chapter by Molyneux in Volume 1 of this series). In sharp contrast to the dismal situation with sleeping sickness, major progress is being made in the control of Chagas disease.

Chagas disease historically has been a major public health problem in endemic countries, which include essentially all of Latin America with the exception of the Caribbean nations. The World Health Organization estimates that at the present time there are 16-18 million persons infected with *T. cruzi*, and essentially all of these individuals will run the risk of developing symptomatic chronic Chagas disease because the infection is life-long and specific drug treatment is little used and lacks efficacy. Even though only 10-30 % of *T. cruzi*-infected persons ever develop chronic symptomatic Chagas disease, the burden of disability and death borne by the endemic countries is enormous. As an example, it was estimated in the early 1990s that in Brazil the yearly cost of medical care, including pacemakers and surgery for megadisease, as well as early pension expenses and time lost by workers due to disability, totaled several billion dollars. Similar calculations done in other endemic countries also have detailed the sizable economic impact of Chagas disease, and it is currently estimated that its total annual cost in all endemic countries is more than US $8 billion. Viewed from a global perspective, Chagas disease represents the third largest tropical disease burden, after malaria and schistosomiasis. This burden will be borne by the affected nations on a continuing basis, as the millions of *T. cruzi*-infected persons gradually develop symptomatic lesions.

In contrast to this bleak picture relating to the prevalence of *T. cruzi* infection and symptomatic Chagas disease, the situation regarding current incidence rates of transmission of the parasite is strikingly brighter. In 1991 the countries of the Southern Cone of South America (Argentina, Brazil, Bolivia, Chile, Paraguay, and Uruguay) undertook an ambitious program aimed at interrupting transmission of *T. cruzi*. The mainstay of the program is the control of the triatomine vectors that transmit the parasite. This is being done through spraying of residual insecticides in houses infested with triatomines, housing improvement to reduce the likelihood of infestation, and education of persons at risk. An additional element in the program is the improvement of serologic screening of donated blood with the goal of eliminating transfusion-associated transmission of *T. cruzi*. During the first eight years of the initiative, the Southern Cone countries invested $340 million in these efforts. All told over two million houses have been sprayed to eliminate vectors and blood screening programs have been intensified on a widespread basis. Now at its midpoint, the initiative has achieved a remarkable level of success. Ongoing surveillance has provided clear evidence that transmission has been interrupted over vast areas in several of the countries in which Chagas disease is endemic. Epidemiologic studies done in selected regions have shown a progressive reduction in prevalence rates in the younger age groups and a gradual reduction in the percent of blood donors who harbor *T. cruzi*. In 1997 Uruguay was certified as being free of transmission, and certification of Chile followed in 1999. Argentina and Brazil are expected to follow suit within several years. Bolivia has the highest prevalence rates of *T. cruzi* infection in the Southern Cone and the initiative has achieved the lowest rates of penetration there. In view of this it may be appropriate to expand programs of chemoprophylaxis of donated blood until effective screening programs can be implemented. It is hoped that complete interruption of *T. cruzi* transmission to humans will be achieved by 2010 throughout the Southern Cone, thus providing enormous medical and social benefits as well as substantive economic rates of return on the monies invested in the program. Programs similar to the Southern Cone initiative were instituted in the Andean countries and in the nations of Central America in 1997, and major progress is being made in these areas as well.

The situation regarding Chagas disease in Mexico merits special mention. Although it has long been recognized as an endemic country, the epidemiology of *T. cruzi* infection in Mexico has not been as extensively characterized as it has in other nations where the illness is a problem. Sporadic reports of both acute and chronic Chagas disease in several regions of Mexico have appeared over the years. Moreover, in a national survey done among blood donors in Mexico without confirmatory testing in the mid-1990s, a 1.5 % overall prevalence rate of *T. cruzi* infection was found, with the highest rates in the states of Hidalgo (2.8 %), Tlaxcala (1.9 %), and Puebla (1.8 %). In a recent study of *T. cruzi* infection among blood donors in the states of Jalisco and neighboring Nayarit, my colleagues and I used screening and confirmatory testing to determine that the prevalence rates of 0.9 % and 0.6 % respectively (unpublished results). In addition, we found that four of nine recipients of blood products from *T. cruzi*-infected donors were infected with the parasite. These findings indicate clearly that *T. cruzi* infection is common among blood donors in Guadalajara and Tepic and that transmission of the organism to recipients of tainted blood is occurring. Serologic testing of blood donors in the study areas should be performed, and the limited epidemiologic evidence available from other areas suggests that testing of donated blood in the rest of Mexico would be appropriate, especially in view of the major internal migrations that have occurred. National blood bank regulations are currently being revised in Mexico and a recommendation for screening for *T. cruzi* is being considered. As far as control of vector-borne transmission is concerned, more epizootiologic and epidemiologic data will have to be obtained in order to focus control measures effeciently in the regions in which *T. cruzi* infection is most common. To date few specific programs for controlling the transmission of *T. cruzi* have been implemented.

Congenital Chagas disease is an issue of major public health concern, as 5-10 % of infants born to *T. cruzi*-infected women are in turn infected. The usefulness of benznidazole or nifurtimox in preventing congenital transmission of *T. cruzi* has not been studied, and in addition the safety of these two drugs in pregnancy has not been established. In this context, then, efforts need to be focused on diagnosing and treating congenital Chagas disease in infants born to mothers with Chagas disease. To this end pregnant women at risk for *T. cruzi* infection should be screened serologically and testing of babies born to positive mothers should follow. Testing infants, however, is problematic because assays for *T. cruzi*-specific IgG would only reflect the maternal infection, and assays for specific IgM antibodies have not been standardized and are not widely available. Parasitologic studies then are the only alternative. The traditional methods of xenodiagnosis and hemoculture are unsatisfactory for congenital Chagas disease, however, in that they lack sensitivity and generally take at least a month to complete. Fortunately, PCR-based assays for detection of parasite DNA in blood obtained from infants, offer a sensitive and specific alternative that can be done quickly. Several laboratories have expended considerable effort in developing such assays, and in my view this approach will become the mainstay of diagnosing congenital *T. cruzi* infection. Unfortunately, despite the fact that all experts in Chagas disease agree that infants and children with congenital Chagas disease should be treated with one of the two drugs available, comprehensive programs for identifying and treating patients with congenital Chagas disease generally have not been implemented.

Another noteworthy issue of public health concern is the possible transmission of *T. cruzi* by organ transplantation. Over the years a small number of such instances have been reported in Latin America. This possibility is particularly worrisome because transplant recipients have a relatively limited ability to control multiplication of the parasites due to the immunosuppression given post-operatively. Earlier reports described transmission of the parasite by transplantation of kidneys and a heart harvested from persons with chronic *T. cruzi* infections. A more poignant instance of transplantation-associated transmission of *T. cruzi* occurred recently in the United States. In this case, several organs were harvested from a *T. cruzi*-infected immigrant from Latin America and all three transplant recipients developed acute Chagas disease. These patients were treated with nifurtimox and it is not yet known if they were cured parasitologically. The problem of *T. cruzi* transmission by organ transplantation could be eliminated by serologic screening of all organ donors in Latin America and those in the United States and other industrialized countries who are at geographic risk for *T. cruzi* infection, and then not transplanting organs from persons found to be infected. Such testing would be difficult to do in a logistical sense, however, and more importantly, given the chronic shortage of organs for transplantation this approach simply is not acceptable. Rather, I suggest screening of all organ donors at risk for *T. cruzi* infection followed by serial serologic, and in symptomatic patients parasitologic, studies in the months following transplantation.

The epidemiology of Chagas disease in the United States and the issue of possible transmission of *T. cruzi* by transfusion and organ transplantation also merit mention. Despite the fact that the sylvatic cycle of *T. cruzi* is present in large areas of the western and southern United States, only five cases of autochthonous transmission to humans have been reported here. Our relatively high housing standards and the low overall vector density are probably the determinants of the rarity of vector-borne transmission of *T. cruzi* to people in the United States. A handful of imported cases of acute Chagas disease have been reported to the Centers for Disease Control and Prevention in the last three decades, but none of these occurred in returning tourists. Even though the number of autochthonous and imported cases of acute Chagas disease may be many times the number reported, the fact remains that the illness is rare in the United States and it does not constitute a major public health concern.

In contrast, the number of persons in the United States with chronic *T. cruzi* infections has increased markedly in recent years. Since 1972 more than 5.5 million persons have emigrated to the United States legally from countries in which Chagas disease is endemic, and several million more have entered illegally. As many as 7 to 8 million of these immigrants may be Mexicans, where as noted above *T. cruzi* infection is widespread, but a sizable percentage has also come from Central America, where *T. cruzi* prevalence is quite high. Over a decade ago my coworkers and I did a study of *T. cruzi* infection among Salvadorans and Nicaraguans in Washington, D.C., and found a 5 % prevalence rate. Studies done in a hospital in Los Angeles where 50 % of donors are Hispanic have shown that between 1:1,000 and 1:500 donors are infected with the parasite. In another study, carried out in seven blood banks in three Southwestern states, approximately 1 in 600 donors with Hispanic surnames was found to harbor *T. cruzi*. In a much larger investigation done in Miami and Los Angeles, the prevalence of *T. cruzi* infection was found to be 1:8,800 in the general donor population and 1:710 in donors who had spent at least a month in an endemic area. From these findings and year 2000 census data relating to both legal and illegal immigrants I estimate that at least 50,000 to 100,000 *T. cruzi*-infected persons now live in the United States. These immigrants pose a risk for transfusion-associated transmission of the parasite in the United States and in other countries to which Latin Americans have emigrated. To date seven such cases have been reported in the United States, Canada, and Europe, all of which occurred in immunosuppressed patients in whom the diagnosis of *T. cruzi* infection was made because of the fulminant course of the illness. Since most transfusions are given to immunocompetent patients in whom acute Chagas disease would be a mild illness, it is reasonable to conclude that many other instances of transfusion-associated transmission of *T. cruzi* have occurred in the United States and other industrialized nations but have not been detected. In the last few years, however, in the United States at least the risk may have been reduced by screening prospective blood donors with questions relating to residence in endemic countries. The question of whether serologic screening of blood donated in the United States for infection with *T. cruzi* has been considered by both public and private entities involved in blood banking for at least a decade. A panel of experts convened recently by the American Red Cross to consider this question unanimously recommended that our blood supply be screened serologically. However, implementation of such a recommendation is made difficult by the fact that the FDA has not cleared any assays for *T. cruzi* infection for use in blood banks.

Finally, it is noteworthy that the enormous progress made to date in reducing vector-borne transmission of *T. cruzi* has been achieved by low-tech, straightforward approaches. Although great strides have been made during the past several decades in understanding the genetics of *T. cruzi*, its interaction with the immune systems of its mammalian hosts, and the pathogenic mechanisms that lead to chronic symptomatic Chagas disease, essentially none of this information has had an impact on the striking success in reducing transmission of the parasite from insects to people. Likewise, the vast knowledge developed has not led to the development of new therapies. These facts may have important implications for the control of other infectious diseases, both in the industrialized nations as well as in developing countries.

Suggested reading

Brener Z, Andrade ZA, Barral-Netto M (eds). 2001. *Trypanosoma cruzi* e Doença de Chagas. Rio de Janeiro: Guanabara Koogan.

Freilij H, Altcheh J. 1995. Congenital Chagas' disease: diagnostic and clinical aspects. Clin Infect Dis 21:551-5.

Gomes ML, Macedo AM, Vago AR, Pena SD, Galvao LM, Chiari E. 1998. *Trypanosoma cruzi*: optimization of polymerase chain reaction for detection in human blood. Exp Parasitol 88:28-33.

Guzman Bracho C, Garcia Garcia L, Floriani Verdugo J, Guerrero Martinez S, Torres Cosme M, Ramirez Melgar C, Velasco Castrejon O. 1998. Risk of transmission of *Trypanosoma cruzi* by blood transfusion in Mexico. Pan Am J Pub Health 4:94-99.

Kirchhoff LV, Gam AA, Gilliam FC. 1987. American trypanosomiasis (Chagas' disease) in Central American immigrants. Am J Med 82:915-20.

Kirchhoff LV, Votava JR, Ochs DE, Moser DR. 1996. Comparison of PCR and microscopic methods for detecting *Trypanosoma*. J Clin Microbiol 34:1171-5.

Kirchhoff LV. 2001. American trypanosomiasis (Chagas disease). In: Principles and Practice of Clinical Parasitology. Gillespie SH, Pearson RD (eds). John Wiley & Sons, West Essex, UK, ppg 335-353.

Leiby DA, Read EJ, Lenes BA, Yund JA, Stumpf RJ, Kirchhoff LV, Dodd RY. 1997. Prevalence of antibodies to *Trypanosoma cruzi*, etiologic agent of Chagas disease, in U.S. blood donors. J Infect Dis 176:1047-52.

Leiby DA, Lenes BA, Tibbals MA, Tames-Olmedo MT. 1999. Prospective evaluation of a patient with *Trypanosoma cruzi* infection transmitted by transfusion. N Engl J Med 341:1237-1239.

Maguire JB, Herwaldt BL, *et al.* Transmission of *Trypanosoma cruzi* by organ transplantation. Morb Mort Wkly Rep (submitted).

Ochs DE, Hnilica V, Moser DR, Smith JH, Kirchhoff LV. 1996. Postmortem diagnosis of autochthonous acute chagasic myocarditis by polymerase chain reaction amplification of a species-specific DNA sequence of *Trypanosoma cruzi*. Am J Trop Med Hyg 34:526-9.

Paredes P, Lomelí-Guerrero A, Paredes-Espinoza M, Ron-Guerrero C, Delgado-Mejía M, Peña-Muñoz J, Kirchhoff LV. Risk of transfusion-associated American trypanosomiasis (Chagas disease) in Mexico: Implications for transfusion medicine in the United States. (submitted)

Schofield CJ, Dias JC. 1999. The Southern Cone initiative against Chagas disease. Adv Parasitol 42:1-27.

INDEX